IV Conference Ia ValSe-Food CYTED and VII Symposium Chia-Link

IV Conference Ia ValSe-Food CYTED and VII Symposium Chia-Link

Editors

Norma Sammán
Mabel Cristina Tomás
Loreto Muñoz
Claudia Monika Haros

Basel • Beijing • Wuhan • Barcelona • Belgrade • Novi Sad • Cluj • Manchester

Editors
Norma Sammán
Centro Interdisciplinario de Investigaciones en Tecnologías y Desarrollo Social para el NOA (CIITED)
CONICET
San Salvador de Jujuy, Argentina

Mabel Cristina Tomás
Centro de Investigación y Desarrollo en Criotecnología de Alimentos (CIDCA)
CONICET-UNLP-CIC
La Plata, Argentina

Loreto Muñoz
Food Science Lab, School of Engineering
Universidad Central de Chile
Santiago, Chile

Claudia Monika Haros
Institute of Agrochemistry and Food Technology (IATA)
Spanish Council for Scientific Research (CSIC)
Valencia, Spain

Editorial Office
MDPI
St. Alban-Anlage 66
4052 Basel, Switzerland

This is a reprint of articles from the Proceedings published online in the open access journal *Biology and Life Sciences Forum* (ISSN 2673-9976) (available at: https://www.mdpi.com/2673-9976/17/1).

For citation purposes, cite each article independently as indicated on the article page online and as indicated below:

Lastname, A.A.; Lastname, B.B. Article Title. *Journal Name* **Year**, *Volume Number*, Page Range.

ISBN 978-3-0365-8836-0 (Hbk)
ISBN 978-3-0365-8837-7 (PDF)
doi.org/10.3390/books978-3-0365-8837-7

Contents

Editorial

Statement of Peer Review

Norma Sammán [1], Mabel Cristina Tomás [2], Loreto Muñoz [3] and Claudia Monika Haros [4,*]

1 Centro Interdisciplinario de Investigaciones en Tecnologías y Desarrollo Social para el NOA (CIITED), Facultad de Ingeniería, Universidad Nacional de Jujuy, Consejo Nacional de Investigaciones Científicas y Técnicas (CONICET), San Salvador de Jujuy 4600, Argentina
2 Centro de Investigación y Desarrollo en Criotecnología de Alimentos (CIDCA), CCT La Plata (CONICET), CICPBA, Facultad de Ciencias Exactas (FCE), Universidad Nacional de La Plata (UNLP), La Plata 1900, Argentina
3 Food Science Lab, School of Engineering, Universidad Central de Chile, Santiago 8330601, Chile
4 Institute of Agrochemistry and Food Technology (IATA), Spanish Council for Scientific Research (CSIC), 46980 Valencia, Spain
* Correspondence: cmharos@iata.csic.es; Tel.: +34-963900022

In submitting conference proceedings to *Biology and Life Sciences Forum*, the volume editors of the proceedings certify that all papers published in this volume have been subjected to peer review administered by the volume editors relative to the publisher. Reviews were conducted by expert referees and held to the professional and scientific standards expected of a proceedings journal:

- Type of peer review: single-blind; double-blind; triple-blind; open; other (please describe): single-blind;
- Conference submission management system: we do not have any system;
- Number of submissions sent for review: all of them (26);
- Number of submissions accepted: all of them (26);
- Acceptance rate (number of submissions accepted/number of submissions received): all accepted (26/26);
- Average number of reviews per paper: three or four (3 or 4);
- Total number of reviewers involved: four (4);
- Any additional information on the review process: no additional comments.

Disclaimer/Publisher's Note: The statements, opinions and data contained in all publications are solely those of the individual author(s) and contributor(s) and not of MDPI and/or the editor(s). MDPI and/or the editor(s) disclaim responsibility for any injury to people or property resulting from any ideas, methods, instructions or products referred to in the content.

Citation: Sammán, N.; Tomás, M.C.; Muñoz, L.; Haros, C.M. Statement of Peer Review. *Biol. Life Sci. Forum* **2022**, *17*, 26. https://doi.org/10.3390/blsf2022017026

Published: 20 December 2022

Editorial

Preface of IV Conference Ia ValSe-Food CYTED and VII Symposium Chia-Link

Norma Sammán [1], Mabel Cristina Tomás [2], Loreto Muñoz [3] and Claudia Monika Haros [4,*]

1 Centro Interdisciplinario de Investigaciones en Tecnologías y Desarrollo Social para el NOA (CIITED), Facultad de Ingeniería, Universidad Nacional de Jujuy, Consejo Nacional de Investigaciones Científicas y Técnicas (CONICET), San Salvador de Jujuy 4600, Argentina
2 Centro de Investigación y Desarrollo en Criotecnología de Alimentos (CIDCA), CCT La Plata (CONICET), CICPBA, Facultad de Ciencias Exactas (FCE), Universidad Nacional de La Plata (UNLP), La Plata 1900, Argentina
3 Food Science Lab, School of Engineering, Universidad Central de Chile, Santiago 8330601, Chile
4 Institute of Agrochemistry and Food Technology (IATA), Spanish Council for Scientific Research (CSIC), 46980 Valencia, Spain
* Correspondence: cmharos@iata.csic.es; Tel.: +34-963900022

We want to walk the full path, with a firm step at each stage, following Machado's verses: "*Walker there is no path, you make the path as you go*".

This book brings together the complete collection of presentations of the IV International Ia ValSe-Food Conference and VII Symposium of the Chia-Link Network, held in the cities of La Plata and Jujuy, Argentina on 14–18 November 2022.

The topics are approached from diverse and articulated perspectives, such as agronomic and nutritional properties, the content of biofunctional compounds, technological characteristics, and the development of new products from ancestral Latin American crops, as a response to the motto of promoting research and innovation in food with valuable Ibero-American seeds.

The presentations are the result of the collaboration and exchange of experiences between the members of the Ia ValSe Group and the Chia-Link Network, which consists of 50 groups including institutions that are a part of the scientific-technological system, researchers, companies, and the industrial/consumer associations from 12 countries.

As is well known, the conservation of biodiversity is essential for food security and nutrition. Nowadays, the world largely relies on a reduced set of staple crops. Consumption of these native foods, along with scientific knowledge, will help to develop local economies and prevent any type of malnutrition and various noncommunicable diseases.

Currently, the complex system of relationships demands interdisciplinary research processes, drawing on a combination of knowledge from different fields without taking into account hierarchies or exclusions.

This book concludes another stage in the process of using our current generation of knowledge to incorporate underexploited native crops with a high content of nutrients into the population's diet. Additionally, this publication strengthened the exchange of information between the participants of the Ia ValSe Group and the Chia-Link Network.

Citation: Sammán, N.; Tomás, M.C.; Muñoz, L.; Haros, C.M. Preface of IV Conference Ia ValSe-Food CYTED and VII Symposium Chia-Link. *Biol. Life Sci. Forum* **2022**, *17*, 27. https://doi.org/10.3390/blsf2022017027

Published: 20 December 2022

Funding: This publication was financed by the Project Ia Valse Food-CYTED (Ref. 119RT0567).

Conflicts of Interest: The authors declare no conflict of interest.

Proceeding Paper

Preparation of Corn Starch Nanoparticles by Wet-Stirred Media Milling for Chia Oil Vehiculization †

María Gabriela Bordón [1,2,3,*], Marcela Lilian Martínez [2,3,4], Fabrizio Chiarini [2], Rodrigo Bruschini [2], Nahuel Camacho [5], Hernán Severini [2], Santiago Daniel Palma [5] and Pablo Daniel Ribotta [1,2,3]

1 Instituto de Ciencia y Tecnología de Alimentos Córdoba (ICYTAC, CONICET-UNC), Córdoba 5000, Argentina
2 Departamento de Química Industrial y Aplicada, Facultad de Ciencias Exactas, Físicasy Naturales, Universidad Nacional de Córdoba (FCEFyN-UNC), Córdoba 5000, Argentina
3 Instituto de Ciencia y Tecnología de Alimentos (ICTA), Facultad de Ciencias Exactas, Físicasy Naturales, Universidad Nacional de Córdoba (FCEFyN-UNC), Córdoba 5000, Argentina
4 Instituto Multidisciplinario de Biología Vegetal (IMBIV, CONICET-UNC), Córdoba 5000, Argentina
5 Unidad de Investigación y Desarrollo en Tecnología Farmacéutica (UNITEFA, CONICET-UNC), Córdoba 5000, Argentina
* Correspondence: maria.gabriela.bordon@mi.unc.edu.ar

† Presented at the IV Conference Ia ValSe-Food CYTED and VII Symposium Chia-Link, La Plata and Jujuy, Argentina, 14–18 November 2022.

Abstract: An organic-free method was applied to produce corn starch nanoparticles, which were designed to stabilize Pickering emulsions containing chia oil, the richest vegetable source of omega-3 fatty acids. The liquid stream resulting from a laboratory-scale mill assisted by zirconia beads was filtered, centrifuged and homogenized to prepare the continuous phase of the emulsions. Experiments were performed as follows: 24 h (milling time), 0.1–0.2 mm (beads' diameter), 1600 rpm (impeller speed), 25% (volume occupied by the grinding media), 1–7% w/v (starch concentration) and 0–1% w/v of sodium dodecyl sulfate (SDS). Particle sizes in the obtained nanosuspensions were reduced from 376–432 nm to 160–200 nm after centrifugation and homogenization. The product formulated with 0.01% w/v of SDS showed the most stable particle size during storage. Hence, this latter formulation was selected to prepare Pickering emulsions. Oil droplets showed surface mean diameters and polydispersity indexes of 283.33 ± 1.53 nm and 1.36 ± 0.03, respectively, with no significant variations during storage for around two weeks. Finally, nanosuspensions containing 7% w/v of starch, and the above three concentrations of SDS, were filtered, centrifuged, homogenized and spray-dried to obtain redispersible powders able to stabilize Pickering emulsions. The most stable particle size after redispersion (385.83 ± 5.85 nm) was obtained with the highest concentration of SDS. Moreover, SEM images revealed the presence of round-shaped particles with sizes below 1 μm. These results highlight that wet-stirred media milling can be applied as a green-method to produce new food-grade starch nanoparticles, which are able to deliver bioactive compounds from chia oil.

Keywords: chia oil; corn starch; nanoparticles; Pickering emulsions; wet-stirred media milling

Citation: Bordón, M.G.; Martínez, M.L.; Chiarini, F.; Bruschini, R.; Camacho, N.; Severini, H.; Palma, S.D.; Ribotta, P.D. Preparation of Corn Starch Nanoparticles by Wet-Stirred Media Milling for Chia Oil Vehiculization. *Biol. Life Sci. Forum* **2022**, *17*, 1. https://doi.org/10.3390/blsf2022017001

Academic Editors: Norma Sammán, Mabel Cristina Tomás, Loreto Muñoz and Claudia Monika Haros

Published: 19 October 2022

1. Introduction

Chia (*Salvia hispanica* L.) oil represents the most abundant vegetable source of omega-3 fatty acids. Despite the health benefits associated with a regular consumption of omega-3-rich oils, the polyunsaturated structure of fatty acids makes this oil highly susceptible to oxidation. Several technological strategies have been applied to stabilize chia oil [1]. Within this context, Pickering emulsions are rising as a very promising alternative for the vehiculization of omega-6 and omega-3 fatty acids in foods [2].

Pickering emulsions are stabilized by micro- or nanoparticles, which are located in the interfacial area of immiscible liquids. For some years, they have attracted the attention of many researchers due to their special characteristics: greater stability, less toxicity, better response to stimuli and resistance to coalescence phenomena [3]. Starch, a very abundant compound in nature, is positioned as a strong candidate for obtaining nanoparticles, which are well recognized within the food sector as novel stabilizing agents for dispersed systems [4].

Nanomilling in the stirred media mills (a top-down technique) is an efficient process for the preparation of ultrafine materials, owing to its advantageous features, viz., ease of operation, simple construction, high size reduction rate, ability to run continuously, and absence of organic solvents [5,6].

The objective of this work was to prepare stable corn starch nanoparticles via wet-stirred media milling, which were further used to stabilize Pickering emulsions containing chia oil. In addition, redispersible powders were obtained by spray-drying of the nanosuspensions due to the ease of storage and transport of solid materials compared with their fluid counterparts.

2. Materials and Methods

2.1. Materials

Chia seed oil (CSO) was extracted from seeds coming from the Salta province (Nutracéutica Sturla SRL, Salta, Argentina), as described by Martínez et al. [7], in a pilot plant screw press (Komet Model CA 59 G, IBG Monforts, Mönchengladbach, Germany). Corn starch was purchased to a local distributor (Distribuidora NICCO, Córdoba, Argentina); sodium dodecyl sulfate (SDS) (Sigma–Aldrich, San Luis, MO, USA) was used as stabilizer during milling experiments. Other reagents were HPLC or analytical grade.

2.2. Milling Experiments and Post-Processing of the Obtained Nanosuspensions

Corn starch nanoparticles (CSN) were prepared by media milling using a NanoDisp® laboratory-scale mill (NanoDisp®, Córdoba, Argentina). The temperature of the process was fixed at 15 °C by circulation of cold water with a Thermo Haake® compact refrigerated circulator (Thermo Fisher Scientific, Waltham, MA, USA) [8].

First, suspensions containing starch (1% w/v) and SDS (0, 0.1 and 1% w/v) were prepared. The resultant mixtures were stirred for 10 min. Afterwards, suspensions and zirconia beads (0.1–0.2 mm and 25% v/v) were placed in the milling chamber and processed at 1600 rpm for 24 h. The obtained nanosuspensions were filtered through a 200 ASTM screen (Zonytest, Buenos Aires, Argentina) (74 μm), and the effects of centrifugation (13,000× g 40 min) and homogenization (18,000 rpm, 2 min, Ultraturrax homogenizer IKA T18, Janke & Kunkel GmbH, Staufen, Germany; followed by 1 cycle at 700 bar, in a high-pressure valve homogenizer, EmulsiFlex C5, Avestin, Ottawa, ON, Canada) on the stability of particle size were analyzed.

2.3. Characterization of the Obtained Nanosuspensions

The average particle size (Z_{av}) and the polydispersity index (PDI) immediately after milling and post-processing, and after storage for one week, were determined by dynamic light scattering at 25 °C (Nano Zetasizer, Malvern Instruments, Worcestershire, UK). The Zeta potential of nanosuspensions were also determined by dynamic light scattering [8].

2.4. Preparation and Characterization of Pickering Emulsions

The filtered and centrifuged nanosuspensions (1% w/v starch and 0.01% w/v SDS) obtained as described in Section 2.2 were blended with CSO (solids/CSO ratio: 5/1 w/w) by high-speed homogenization (18,000 rpm, 2 min). Subsequently, the coarse emulsions were processed in a high-pressure valve homogenizer (2 cycles, 700 bar, EmulsiFlex C5, Avestin, Ottawa, ON, Canada) to obtain fine emulsions. The oil droplet size distribution and polydispersity index were determined according to [1] with a LA 950V2 Horiba (Kyoto, Japan) analyzer, and the Zeta potential was measured according to Section 2.3.

2.5. Spray-Drying of Nanosuspensions and Characterization of Powders

Nanosuspensions containing 7% w/v of starch, and the three concentrations of SDS (Section 2.2), were filtered, centrifuged, homogenized (Sections 2.2 and 2.4) and spray-dried to obtain redispersible powders able to stabilize Pickering emulsions. The spray-drying process was carried out in a laboratory-scale spray-dryer, Büchi B-290 (Büchi Labortechnik AG, Flawil, Switzerland) equipped with a two-fluid nozzle atomizer. The drying conditions given by Fu et al. [9] were followed. The obtained powders were redispersed in Milli-Q water to determine the Zav the and PDI as described in Section 2.3. Finally, the morphology of powders was evaluated by scanning electron microscopy (SEM, LSM5 Pascal; Zeiss, Oberkochen, Germany).

3. Results and Discussion

3.1. Characterization of the Obtained Nanosuspensions and Pickering Emulsions

Nanomilling refers to the reduction in particle size below 1000 nm by wet media-milling, and the intermediate product is a nanoparticle suspension or nanosuspension [5]. In order to obtain a stable product, the selection of the stabilizer formulation is a resource-demanding task, with potentially serious consequences such as aggregation, Ostwald ripening, sedimentation of particles, and cake formation during milling and storage [10]. In the present study, a widely-used amphiphilic surfactant such as SDS was used to provide electrostatic stabilization.

Milling could modify the granular morphology and micro molecular structure in starch, leading to improved physicochemical properties such as low gelatinization temperature and paste viscosity, enhanced redispersion, and large specific surface area, thus widening the scope of starch applications [6]. Table 1 shows the average particle size and polydispersity index of fresh nanosuspensions and after storage for one week. The values for different concentrations of SDS, and through different processing steps (milling, centrifugation and homogenization), are included. The initial average particle sizes of starch suspensions, before milling, were around 15 μm (1% starch), 14 μm (1% starch + 0.01% SDS) and 90.6 μm (1% starch + 1% SDS). After a similar milling time (25 h), with native corn starch granules ranging from 2–20 μm, Lu et al. [6] observed Z_{av} values around 700 nm, above those found in Table 1.

Table 1. Average particle size and polydispersity index of starch nanoparticles.

Formulation-Process	$Z_{av,0}$ (nm)	$Z_{av,f}$ (nm)	PDI_0	PDI_f
1% S-M	416.13gh	2194.33f	0.332def	0.275ab
1% S-C	212.43c	300.73bc	0.233ab	0.383de
1% S-H	315.40e	502.87e	0.184a	0.504f
1% S-C+H	199.93bc	274.53b	0.260bc	0.352cd
1% S + 0.01% SDS-M	444.00h	416.90d	0.314cde	0.397de
1% S + 0.01% SDS-C	182.87ab	182.37a	0.247abc	0.242a
1% S + 0.01% SDS-H	373.60f	341.93c	0.365efg	0.276ab
1% S + 0.01% SDS-C+H	161.27a	158.10a	0.239ab	0.228a
1% S + 1% SDS-M	397.57fg	468.73de	0.400g	0.448ef
1% S + 1% SDS-C	265.23d	247.47b	0.288bcd	0.273ab
1% S + 1% SDS-H	407.97g	341.97c	0.387fg	0.291abc
1% S + 1% SDS-C+H	265.43d	253.33b	0.380efg	0.328bcd

S: starch; M: milling; C: centrifugation; H: homogenization; "0" and "f" subscripts indicate initial and final values after one week, respectively; different letters in each column indicate statistically significant differences ($p < 0.05$) among samples.

The significantly higher ($p < 0.05$) average particle size, observed with the highest concentration of surfactant, may be attributed to the formation and aggregation of SDS micelles above its critical micelle concentration (0.23% w/v, 25 °C in the absence of any other additive). Moreover, it has been suggested that beyond a certain level, the use of a higher surfactant concentration may not prevent particle aggregation, but promote Ostwald ripening [10]. Multiple range tests by formulation show significantly ($p < 0.05$) higher Z_{av} and PDI values with 1% SDS. On the other hand, the lowest values were observed with 0.01% SDS, suggesting a proper stabilization of nanoparticles despite the lesser surface charge given by the Zeta potential (Table 2). Therefore, this latter surfactant concentration was selected for the forthcoming formulation of Pickering emulsions.

Table 2. Zeta potential of starch nanoparticles.

Formulation-Process	Zeta Potential (mV)
1% S-M	−8.70ef
1% S-C	−7.68f
1% S-H	−9.59ef
1% S-C+H	−16.70d
1% S + 0.01% SDS-M	−9.23ef
1% S + 0.01% SDS-C	−12.43e
1% S + 0.01% SDS-H	−20.97c
1% S + 0.01% SDS-C+H	−19.00cd
1% S + 1% SDS-M	−62.27a
1% S + 1% SDS-C	−41.47b
1% S + 1% SDS-H	−38.70b
1% S + 1% SDS-C+H	−39.40b

S: starch; M: milling; C: centrifugation; H: homogenization; different letters in each column indicate statistically significant differences ($p < 0.05$) among samples.

Additional multiple range tests showed a significant ($p < 0.05$) reduction in Z_{av} and PDI after centrifugation, highlighting a necessary removal of large suspended particles. The homogenization of milled suspensions may further promote the disruption of particles [1], as can be seen from Z_{av} values. However, no additional reduction in Z_{av} and PDI values were observed ($p > 0.05$) after homogenization of the centrifuged nanosuspensions, evidencing a high mechanical input necessary to disrupt such nanoparticles. The homogenization step is key during emulsification of oil droplets, though. Hence, it was included as part of the preparation of Pickering emulsions.

The oil droplet size distribution of fresh emulsions, and after storage for one week, is given in Table 3 and shown in Figure 1. The Zeta potential values are shown in Table 3. As can be seen, emulsions containing only native starch or SDS were also prepared. It has been informed that native starch granules may not form stable Pickering emulsions [2]. Indeed, sedimentation of solid particles was observed during storage (images not shown). Decreasing starch particle size tends to decrease the oil droplet size, while increasing the storage stability [2]. Moreover, the adsorption behavior of solid particles is different from that of small-molecule surfactants. Slow adsorption of soluble starch particles enables their rearrangement at the oil/water interface. Once these particles are adsorbed, irreversible dense interfacial layers may be formed due to the high-energy barrier against desorption [4], which might explain the stability observed in the emulsions of the present study, despite their low surface charge (−18.43 mV). Finally, the highest surface charge (−48.93 mV) was recorded for the SDS-stabilized emulsions, accounting for their stability during storage.

Table 3. Oil droplet size distribution and zeta potential of emulsions.

Emulsion	Zeta Potential (mV)	$D_{32.0}$ (μm)	$D_{32\cdot f}$ (μm)	PDI_0	PDI_f
Native starch	−8.35c	8.83a	12.29a	3.84a	60.86a
1% milled starch + 0.01% SDS	−18.43b	0.28c	0.47c	1.36b	1.73b
0.01% SDS	−48.93a	0.74b	4.29b	1.23c	1.22c

D_{32}: Sauter mean diameter; different letters in each column indicate statistically significant differences ($p < 0.05$) among samples.

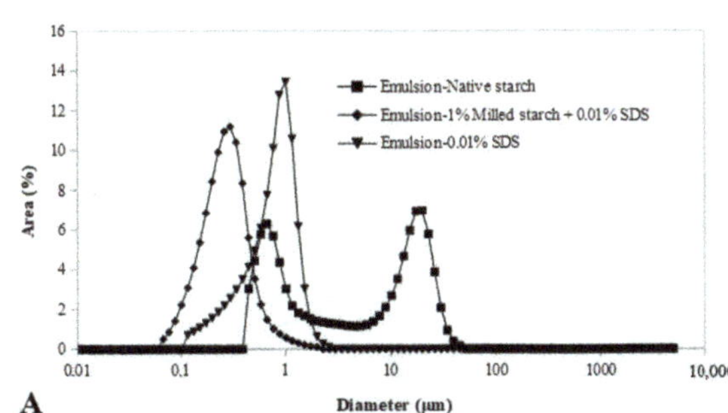

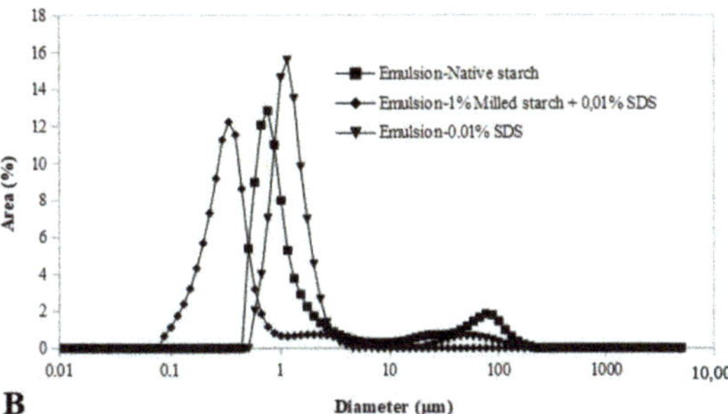

Figure 1. Oil droplet size distribution in emulsions prepared with native starch, 1% milled starch + 0.01% SDS, and 0.01% SDS. (**A**) Fresh emulsions; (**B**) after storage for one week.

3.2. Characterization of the Powders Obtained by Spray-Drying

Redispersible powders, meant to stabilize Pickering emulsions, were obtained by spray-drying of nanosuspensions containing 7% *w*/*v* of starch, and the three concentrations of SDS. The average particle size and polydispersity index, before and after spray-drying, are given in Table 4. Moreover, SEM micrographs of powders (including native starch and SDS) are shown in Figure 2. According to Table 4, SDS greatly improves ($p < 0.05$) the dispersion of powders, as reflected by the highest Z_{av} (881.93 nm) and PDI (1.00) values when no surfactant was used. When a suspension is atomized, a new air/water interface is formed [1] and rapidly stabilized by small-molecule surfactants such as SDS. Regarding the particles' morphology, corn starch granules (Figure 1A,B) showed the characteristic truncated shape of native starch [9]. As a consequence of the milling operation, the internal structure of granules was modified and the resulting particles were swollen considerably [2,9]. With subsequent spray-drying, the volumetric shrinkage was remarkable (Figure 2E–J). Finally, it should be pointed out that the development of wrinkled surface morphologies in the gelatinized starch particles after milling might follow the same mechanisms found in solutions of polymeric macromolecules, such as maltodextrins and whey proteins. The effective moisture diffusivity in such solutions is very low, which results in high moisture gradients between drying air and the interior of droplets/particles. In order to reduce the diffusion path, and to provide a greater interfacial area for moisture evaporation, the drying of droplets try various ways to reduce the distance between the particle surface and its core. This ultimately leads to the deviation from sphericity, and to the formation of surface folds and troughs that facilitates the outward diffusion of water [9].

Table 4. Average particle size and polydispersity index of nanosuspensions and powders obtained after spray-drying (7% *w*/*v* starch).

Formulation-Process	$Z_{av,0}$ (nm)	$Z_{av, powder}$ (nm)	PDI_0	PDI_{powder}
7% S-C+H	469.37c	881.93b	0.31b	1.00b
7%S + 0.07%SDS-C+H	329.40a	200.50a	0.25a	0.29a
7%S + 1%SDS-C+H	359.30b	282.67a	0.24a	0.27a

S: starch; C: centrifugation; H: homogenization; different letters in each column indicate statistically significant differences ($p < 0.05$) among samples.

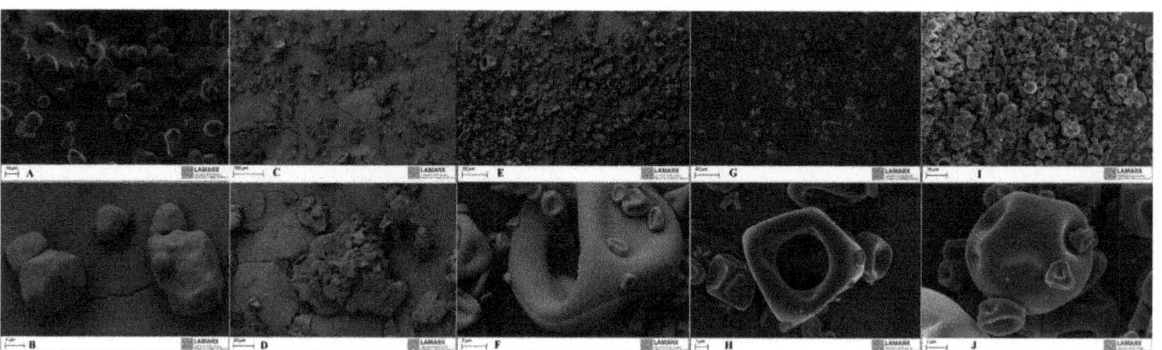

Figure 2. SEM micrographs of powders. (**A**,**B**) native corn starch; (**C**,**D**) SDS; (**E**,**F**) 7% milled starch; (**G**,**H**) 7% milled starch + 0.07% SDS; (**I**,**J**) 7% milled starch + 1% SDS.

4. Conclusions

The stability of particle size in the starch nanosuspensions proved to be greatly affected by the presence of surfactants and post-processing after milling (mainly centrifugation and homogenization). Subsequently, the stability of oil droplets during storage for one week in Pickering emulsions could be confirmed, and might be attributed to the slow adsorption and rearrangement of soluble starch particles at the oil/water interface. This study highlights that wet-stirred media milling can be applied as a green-method to produce new food-grade starch nanoparticles, which are able to deliver bioactive compounds from chia oil.

Author Contributions: Conceptualization, M.G.B., M.L.M., P.D.R. and S.D.P.; methodology, M.G.B., F.C., R.B. and N.C.; software, M.G.B., F.C. and R.B.; validation, M.G.B., F.C., R.B. and N.C.; formal analysis, M.G.B., M.L.M., P.D.R. and S.D.P.; investigation, M.G.B., F.C., R.B. and N.C.; resources, M.L.M., P.D.R., S.D.P. and H.S.; data curation, M.G.B., F.C., R.B. and N.C.; writing—original draft preparation, M.G.B., F.C. and R.B.; writing—review and editing, M.G.B., F.C. and R.B.; visualization, M.G.B., F.C. and R.B.; supervision, M.L.M., P.D.R., S.D.P. and H.S.; project administration, M.L.M., P.D.R., S.D.P. and H.S.; funding acquisition, M.L.M., P.D.R. and S.D.P. All authors have read and agreed to the published version of the manuscript.

Funding: This work was supported by grant Ia ValSe-Food-CYTED (119RT0567), PICT-2019-01122 and SECyT–UNC.

Institutional Review Board Statement: Not applicable.

Informed Consent Statement: Not applicable.

Data Availability Statement: Not applicable.

Conflicts of Interest: The authors declare no conflict of interest.

References

1. Bordón, M.G.; Paredes, A.J.; Camacho, N.M.; Penci, M.C.; González, A.; Palma, S.D.; Ribotta, P.D.; Martínez, M.L. Formulation, spray-drying and physicochemical characterization of functional powders loaded with chia seed oil and prepared by complex coacervation. *Powder Technol.* **2021**, *391*, 479–493. [CrossRef]
2. Zhu, F. Starch based Pickering emulsions: Fabrication, properties, and applications. *Trends Food Sci. Technol.* **2019**, *85*, 129–137. [CrossRef]
3. Gonzalez-Ortiz, D.; Pochat-Bohatier, C.; Cambedouzou, J.; Bechelany, M.; Miele, P. Current trends in Pickering emulsions: Particle morphology and applications. *Engineering* **2020**, *6*, 468–482. [CrossRef]
4. Xu, T.; Yang, J.; Hua, S.; Hong, Y.; Gu, Z.; Cheng, L.; Li, Z.; Li, C. Characteristics of starch-based Pickering emulsions from the interface perspective. *Trends Food Sci. Technol.* **2020**, *105*, 334–346. [CrossRef]
5. Breitung-Faes, S.; Kwade, A. Mill, material, and process parameters-A mechanistic model for the set-up of wet-stirred media milling processes. *Adv. Powder Technol.* **2019**, *30*, 1425–1433. [CrossRef]
6. Lu, X.; Xiao, J.; Huang, Q. Pickering emulsions stabilized by media-milled starch particles. *Food Res. Int.* **2018**, *105*, 140–149. [CrossRef] [PubMed]
7. Martínez, M.L.; Marín, M.A.; Faller, C.M.S.; Revol, J.; Penci, M.C.; Ribotta, P.D. Chia (*Salvia hispanica* L.) oil extraction: Study of processing parameters. *LWT Food Sci. Technol.* **2012**, *47*, 78–82. [CrossRef]
8. Paredes, A.J.; Camacho, N.M.; Schofs, L.; Dib, A.; Zarazaga, M.d.P.; Litterio, N.; Allemandi, D.A.; Sánchez Bruni, S.; Lanusse, C.; Palma, S.D. Ricobendazole nanocrystals obtained by media milling and spray drying: Pharmacokinetic comparison with the micronized form of the drug. *Int. J. Pharm.* **2020**, *585*, 119501. [CrossRef] [PubMed]
9. Fu, Z.-Q.; Wang, L.-J.; Li, D.; Adhikari, B. Effects of partial gelatinization on structure and thermal properties of corn starch after spray drying. *Carbohydr. Polym.* **2012**, *88*, 1319–1325. [CrossRef]
10. Li, M.; Azad, M.; Davé, R.; Bilgili, E. Nanomilling of drugs for bioavailability enhancement: A holistic formulation-process perspective. *Pharmaceutics* **2016**, *8*, 17. [CrossRef] [PubMed]

Proceeding Paper

Amaranth and Chia: Two Strategic Ancestral Grains for Mexico's Sustainability †

Francisco Valenzuela Zamudio and Maira Rubi Segura Campos *

Facultad de Ingeniería Química, Universidad Autónoma de Yucatán, Periférico Norte Km. 33.5, Tablaje Catastral 13615, Colonia Chuburná de Hidalgo Inn, Mérida 97203, Mexico
* Correspondence: maira.segura@correo.uady.mx
† Presented at the IV Conference Ia ValSe-Food CYTED and VII Symposium Chia-Link, La Plata and Jujuy, Argentina, 14–18 November 2022.

Abstract: The demand for food worldwide due to population growth, rising hunger and malnutrition, the adverse effects of climate change, the overexploitation of natural resources, biodiversity loss, and food loss and waste are severe challenges faced by global food and agricultural production systems. A transition to sustainable food and agriculture that ensures global food security and provides economic and social opportunities, while protecting the ecosystem, must occur to achieve sustainability. Due to their numerous nutritional, agronomical, and health advantages, amaranth and chia are attractive grains for sustainable development and product patenting. Nonetheless, there is still a lack of a legal framework and relevant policies to encourage their cultivation. The objective of this study was to summarize the current production state, intellectual property, and legal and political frameworks of both grains. A revision of the most recent literature on the therapeutic potential and nutrient composition of amaranth and chia was conducted. The current production, production value data, intellectual property, legal framework, and policy information were retrieved from public consultation websites and then analyzed. The results of this revision indicate that the production of amaranth and chia grains significantly increased between 2010 and 2016, along with the publication of intellectual property patent applications. Chia and chia-derived products are not regulated by any current law, while amaranth has received more political attention due to its positive results in production numbers. More evidence of the sustainable features of both grains, along with a political framework that encourages their cultivation, may help Mexico to achieve food sustainability.

Keywords: amaranth; chia; sustainability; legal framework; intellectual property

Citation: Valenzuela Zamudio, F.; Segura Campos, M.R. Amaranth and Chia: Two Strategic Ancestral Grains for Mexico's Sustainability. *Biol. Life Sci. Forum* **2022**, 17, 2. https://doi.org/10.3390/blsf2022017002

Academic Editors: Norma Sammán, Mabel Cristina Tomás, Loreto Muñoz and Claudia Monika Haros

Published: 19 October 2022

1. Introduction

The agriculture sector is the world's largest employer, providing income for 40 percent of the world's population today. It is the largest source of income and employment for poor rural households and provides up to 80 percent of the food that is consumed globally. Nevertheless, up to 811 million people suffer from hunger, and the number of people facing acute food insecurity has more than doubled from 135 million to 276 million since 2019. For this reason, global organizations such as the United Nations have been promoting several lines of action to reduce world hunger and food insecurity with a sustainability approach, simultaneously tackling other issues such as environmental protection, health and social protection, and job opportunities. To that end, researchers have investigated several food sources that ensure sustainable production systems that help maintain ecosystems and have strengthened the capacity for adaptation to climate change and extreme weather, in addition to providing incomes for small-scale food producers, particularly women and indigenous peoples [1]. There is evidence that two Mexican native grains have the potential to meet these criteria: the pseudo-cereal grain, amaranth, and the oleaginous chia. Both ancient grains offer outstanding nutritional value with a high protein content (14.5–24%)

and numerous health benefits, such as antidiabetic, antihypertensive, antiobesity, and antioxidative activities [2]. Nonetheless, even though both these grains have proven to be more profitable than others, such as maize, and although there is interest in protecting the intellectual property of the technology related to cultivating these grains, amaranth and chia cultivation are not successfully promoted in Mexico. The relevant legal frameworks and policies are not well established or have not been revised. This study aims to summarize the current federal policies and intellectual protection tendencies of amaranth and chia in Mexico from 2000 to the present in 2022.

2. Methods

2.1. Production Data

All production and production value data were retrieved using the public consultation software "Agroalimentary Information System for Consultation" (SIACON) produced by the "Agri-food and Fisheries Information Service" (SIAP). The search parameters were the following: all federative states of Mexico were selected, only amaranth and chia data were included, the agricultural cycle was set to include all seasons, with irrigation methods and rainfed systems, and the consulting period was limited to the period from the year 2000 to the end of 2021. The selected variables were the total production in tons and production value in thousands of pesos.

2.2. Intellectual Property Information

Intellectual property information regarding amaranth and chia-related technologies was retrieved from the Industrial Property Gazette Information System (SIGA) (https://siga.impi.gob.mx/newSIGA/content/common/principal.jsf (accessed on 1 July 2022)) by the Mexican Institute of Industrial Property (IMPI) using the specialized search tool. The search parameters were the following: only utility patents were consulted, and the keywords used were "amaranth" and "chia," using quotation marks to retrieve the exact term. There were no restrictions on the time of publication. Missing information on international patents was completed by searching the patent ID on the Espacenet website (https://worldwide.espacenet.com/ (accessed on 1 July 2022)) of the European Patent Organization. Data were analyzed using Tableau Public (https://public.tableau.com/en-us/s/ (accessed on 15 July 2022)).

2.3. Legal Framework

Current legislation information was retrieved by consulting the norms catalog of the "Comprehensive Standards and Conformity Assessment System" (SINEC) webpage (https://www.sinec.gob.mx/SINEC/Vista/Normalizacion/BusquedaNormas.xhtml (accessed on 19 July 2022)) using the keywords "amaranth" and "chia." Other legal information was retrieved from the Legislation Chamber website (http://www3.diputados.gob.mx/camara/001_diputados/ (accessed on 19 July 2022)).

3. Results and Discussion

3.1. Sustainable Value of Amaranth and Chia

The global food and agricultural production systems are facing severe challenges, including the demand for food due to population growth, rising hunger and malnutrition, the adverse effects of climate change, the overexploitation of natural resources, biodiversity loss, and food waste. The existing food and agricultural systems have failed to meet the world's food requirements, and more than 811 million people continue to suffer from hunger and malnutrition. There is an urgent need to accelerate a transition to sustainable food and agriculture that will ensure global food security and provide economic and social opportunities while protecting the ecosystems on which agriculture depends. For agriculture and food systems to be sustainable, they must benefit the people who work on the land; they must use novel technologies, markets should be accessible, and they must provide appropriate employment opportunities. Hence, the most fundamental task for the scientific community in this matter is to research sustainable food systems that meet these sustainability criteria or to provide enough evidence for the encouragement of the extensive production of existing sustainable alternatives to conventional food systems [1]. Amaranth and chia are known for their potential to contribute to agrobiodiversity, environmental sustainability, and welfare in food and foodstuff production worldwide. Native to southern Mexico and Central America, these crops are popular as highly profitable grains because of their high nutritional value, fast growth, tolerance of extreme conditions, and ability to grow in poor soils. Amaranth and chia's ability to adapt to adverse growing conditions, such as nutrient-poor soils and wide temperature ranges, make them attractive options as sustainable crops in semi-arid regions. In pre-Columbian civilizations, amaranth and chia formed the basis of human nutrition; the Aztec and Mayan civilizations used amaranth and chia for dietary and other cultural purposes. The nutrient composition of amaranth and chia varies, depending on growing conditions; both grains offer high nutritional value and have a generally positive impact on human health. Amaranth is a gluten-free pseudo-grain that is ideal for people with celiac disease; it is becoming an increasingly popular superfood because of its high-quality but low content of carbohydrates (3%), fiber (4%), lipids (7%), essential amino acids (5.85%), squalene, tocopherols, phenolic compounds, flavonoids, phytic acid, vitamins, and minerals. Amaranth also contains unsaturated fatty acids, such as linoleic acid (omega-6 fatty acid) (2.736%) and alpha-linolenic acid (omega-3 fatty acid) (0.042%), along with several antioxidants. Chia seeds are characterized by their low carbohydrate content (3.4%) and high protein (18.9%), and fat (31.2%) contents. They are also high in omega-3 (19.5%) and omega-6 (5.2%) fatty acids. Both grains' high concentrations of high-quality nutrients and bioactive compounds are associated with a decreased risk of coronary heart disease, hypertension, type 2 diabetes, rheumatoid arthritis, autoimmune diseases, and cancer [2]. Both grains are slowly being incorporated into the diet to improve human health. Consequently, the production, consumption, and demand for chia and amaranth have generally increased in Mexico and worldwide.

3.2. National Production and Production Value of Amaranth and Chia

In the last two decades, the total production of amaranth and chia shared similar behaviors, having a breakthrough in the same period (2010–2015). Amaranth production (Figure 1A) from 2000 to the end of 2010 was consistent, with an average output of around 3600 tons per year, and then exponentially increased, peaking in 2015 (8551 tons). The production subsequently decreased, experiencing a 27.75% drop in 2021 from the total output in 2015. On the other hand, chia production (Figure 1B) was not significant before 2010. In 2013, the total production of chia exponentially increased, peaking in 2014 (9548 tons), surpassing the amaranth production in the same year and the maximum historical output of amaranth in 2015. Nonetheless, from 2016 to the end of 2019, the total production significantly decreased, dropping 37.6%. The last three years of production slightly increased, reaching 4771 tons in 2021, which was still less than the amaranth production. Generally, the production values of these grains show that they are both quite profitable (Figure 1). The maximum production value of amaranth in the last decade was achieved in 2015, reaching over 95 million pesos; the production value increased in 2013 and decreased between 2016 and 2019. In 2021 a growth tendency was observable. The production value of chia is far superior. In 2014, that value reached a total of 420,701,810 pesos and, as was the case with the total production of the grain, along with the production value of amaranth, it decreased between 2016 and 2019, with an observable growth tendency in 2021. In profit terms, using the year 2015 as a reference when the total production and the production value of both grains were above average, the value per ton of amaranth produced was 11,115 pesos, while a ton of chia had a value of 46,628 pesos. In comparison, the most commonly grown grain in Mexico, maize, had a value per ton of 3,423 pesos in the same year. In contrast, using the same year as a reference (2015), in 2021, the value of a ton of maize had increased by 57%, while amaranth and chia had increased by 4% and 22.4%, respectively. Whereas the value production of amaranth and chia is not as high as it was five years ago, the Mexican agriculture industry, in general, is growing, reaching 756,666,728,000 pesos in 2021 [3]. However, inflation should be considered as a factor in such a value.

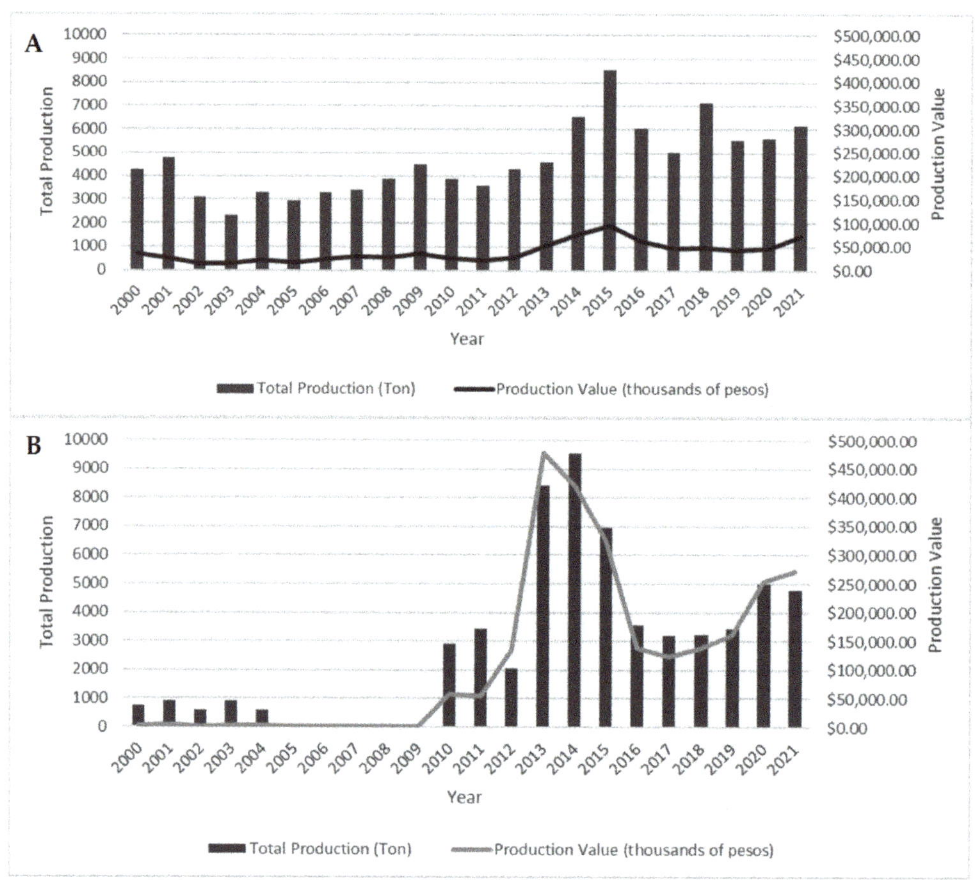

Figure 1. Total production and production value, from 2000 to 2021, of amaranth (**A**) and chia (**B**) in Mexico.

3.3. Intellectual Property Rights of Amaranth and Chia-Based Products and Processes

Amaranth-related technologies surfaced in Mexico in 1994 when the first amaranth-related patent was published in the Mexican Institute of Intellectual Property (IMPI). Ever since then, a total of ten patents and twenty-seven applications for patents have been published in the IMPI gazette (Figure 2A). The year 2014 marked the most active year in intellectual property behavior in the field. On the other hand, interest in patenting chia-related technology started in 2000 (Figure 2B), when the first application was published. Since then, only four patents have been conceded, and thirteen applications have been solicited, marking 2018 as the year with the most applications.

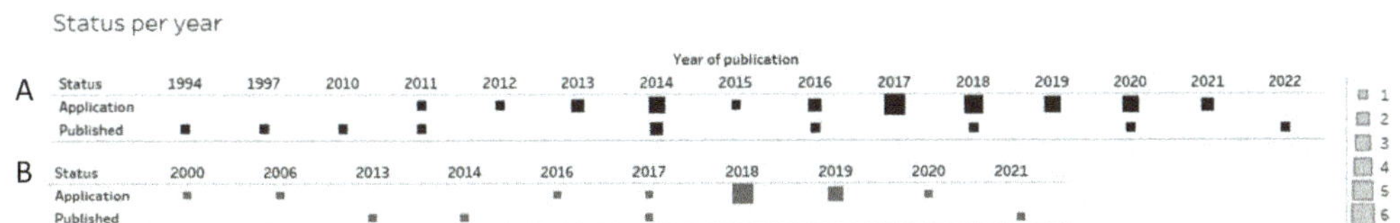

Figure 2. Publication status of amaranth (**A**) and chia-related (**B**) patents through the years.

In terms of technologies, the International Patent Classification (IPC) technology is a fast way to analyze the most common technologies that are yet to be patented. In this context, most of the amaranth-related technology fits the category of "Foods, foodstuffs, or non-alcoholic beverages," the most frequently referenced technology refers to their preparation or treatment (e.g., cooking, the modification of nutritive qualities, and physical treatment), the preservation of foods or foodstuffs, modifying the nutritional qualities of foods, and dietetics (Figure 3A). To a lesser extent, amaranth-related patents also include those related to fermentation, mixing, and transportation technologies. The technologies for chia also are mostly related to food type. As with amaranth-related technologies, the most repeated technology refers to the preparation or treatment preparations for foods, foodstuffs, or non-alcoholic beverages, including the modification of nutritive qualities (Figure 3B). There is also technology related to medical, dental, or toilet applications. There are no patent publications related to grain-mixing, extraction, and transportation technologies. The word-cloud analysis of titles regarding amaranth and chia patents (Figure 4) gives a general overview of the most repeated technologies. In both cases, food terms stand out, such as the words "functional" (functional), "bebidas" (beverages), "harina" (flour), "semillas" (grains), and "nutraceútico" (nutraceuticals), among many others. An important feature to notice is that many words for biological activities appear in both clouds, such as antihypertensive, diabetes, probiotic, and obesity [4]. There is a clear tendency in the intellectual property of both grains toward the positive health features of amaranth and chia.

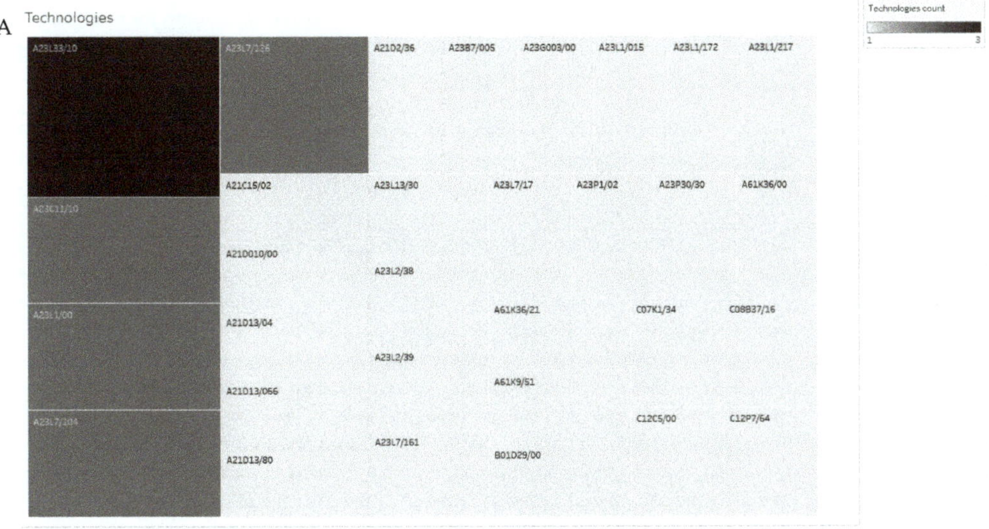

Figure 3. *Cont.*

Figure 3. Treemap of the International Patent Classification (IPC) of amaranth (**A**) and chia-related (**B**) patents. A01G (horticulture; cultivation of vegetables, flowers, rice, fruit, vines, hops or seaweed; forestry; watering); A21C (machines or equipment for making or processing doughs; handling baked articles made from dough); A21D (treatment, such as the preservation, of flour or dough for baking, e.g., by the addition of materials; baking; bakery products; the preservation thereof); A23B (preserving, e.g., by canning, meat, fish, eggs, fruit, vegetables, edible seeds; chemical ripening of fruit or vegetables; the preserved, ripened, or canned products); A23C (dairy products, e.g., milk, butter or cheese; milk or cheese substitutes; the making thereof (obtaining protein compositions for foodstuffs)); A23D (edible oils or fats, e.g., margarines, shortenings, cooking oils (obtaining, refining, preserving; hydrogenation); A23G (cocoa; cocoa products, e.g., chocolate; substitutes for cocoa or cocoa products; confectionery; chewing gum; ice-cream; the preparation thereof); A23K (feeds specially adapted for animals; methods specially adapted for the production thereof); A23L (foods, foodstuffs, or non-alcoholic beverages; their preparation or treatment, e.g., cooking, the modification of nutritive qualities, physical treatment (shaping or working), the preservation of foods or foodstuffs, in general); A23P shaping or working of foodstuffs); A61K (preparations for medical, dental, or toilet purposes (devices or methods specially adapted for bringing pharmaceutical products into particular physical or when administering forms; chemical aspects of, or use of materials for the deodorization of air, for disinfection or sterilization, or for bandages, dressings, absorbent pads or surgical articles; soap compositions); A61Q (specific use of cosmetics or similar toilet preparations); B01D (separation (separating solids from solids by wet methods, using pneumatic jigs or tables, by other dry methods; magnetic or electrostatic separation of solid materials from solid materials or fluids, separation by high-voltage electric fields; centrifuges; vortex apparatus; presses intended for the squeezing-out of liquid from liquid-containing materials); C07K (peptides containing β-lactam rings; cyclic dipeptides not having in their molecule any other peptide link than those which form their ring, e.g., piperazine-2,5-diones; ergot alkaloids of the cyclic peptide type; single cell proteins, enzymes; genetic engineering processes for obtaining peptides); C08B (polysaccharides; derivatives thereof (polysaccharides containing less than six saccharide radicals attached to each other by glycosidic linkages; fermentation or enzyme-using processes; production of cellulose); C08L (compositions of macromolecular compounds (compositions based on polymerizable monomers; artificial filaments or fibers); textile treatment compositions); C12C (beer; preparation of beer by fermentation; aging or ripening of beer by storing; methods for reducing the alcohol content after fermentation; methods for increasing the alcohol content after fermentation; venting devices for casks, barrels, or the like; the preparation of malt for making beer; preparation of hops for making beer); C12P (fermentation or enzyme-using processes to synthesize a desired chemical compound or composition or to separate optical isomers from a racemic mixture).

Figure 4. Word-cloud analysis of the titles of amaranth and chia-related patents.

3.4. *The Legal Framework of Chia and Amaranth Production*

Legal regulations in Mexico include norms that are mandatory (NOM) and norms that are not mandatory (NMX); the latter are published in the "Official Journal of the Federation" (DOF), which is the leading benchmark of legality and the point of reference for all regulations in the country. Within this legal framework, two NOMs regulating amaranth are currently in effect and were redacted by the Regulations Secretariat of Commerce and Industrial Development (SCFI); neither NOM has an international equivalent. The first NOM that was published is the NMX-FF-114-SCFI-2009, which regulates the amaranth plant (*Amaranthus* spp.) for human use and consumption, giving specifications and test methods. The other, NMX-FF-116-SCFI-2010, regulates the products derived from the amaranth plant [5,6]. There are no NMX or NOM for regulating chia seeds or the products derived from them. An initiative was established for the publication of The Mexican Amaranth Promotion Law, which was published in April 2021 in the legislative chamber and aimed to lay the foundations for linking and organizing the production of amaranth, thereby promoting its integral development, implementing strategies that stimulate the advancement and growth of the national output, keeping the advocacy for sustainability goals. It also looks to consolidate the amaranth production chain in Mexico as a financial sector with worldwide leadership in science and technology [7]. In this context, the Operating Rules of the Production for Well-Being Program of the Ministry of Agriculture and Rural Development, in accordance with the National Development Plan (PND) 2019–2024 and objective number two of the Sustainable Development Goals of the UN, establishes strategies to achieve self-sufficiency in the basic foodstuffs consumed by the population. In this sense, food self-sufficiency must be seen in three dimensions: producing the food consumed, generating the inputs and elements required for food production, and developing the necessary knowledge to increase production, effectively meeting the dietary needs of the current and future population. To achieve this, the program has one specific objective, which is to provide liquidity to small and medium-scale producers, preferably of grains, including amaranth and chia, wherein small- and medium-scale producers of amaranth or chia, farming 2 to 20 hectares, receive a per-hectare support fee of up to 3000 pesos. In comparison, other grains such as corn, beans, and rice receive a support fee of 1200 pesos per hectare [8].

3.5. Future Perspectives

Most of the commercial interest in the cultivation of amaranth started with the first national congress of amaranth producers, celebrated in 2015; in consequence, amaranth production is present in legislative conversations, and, in 2017, it was presented as part of a proposal for the management of non-communicable chronic diseases, such as type 2 diabetes, the second leading cause of death in Mexico. Later, in 2018, Mexico hosted the first international congress of amaranth [9]. It is important to note that in the same year that the first congress of amaranth was held (2015), amaranth production reached the highest level of tons produced and production value. In subsequent years, the number of applications for the intellectual protection of amaranth-related technology increased, setting a precedent for the importance of having a political and legal framework to boost the agricultural production of grains, the products derived from such grains, and the innovation of technology to achieve it. This is not the case for chia, where there is a practically nonspecific legal framework related to this grain's production or product derivatives. However, the production value of the grain and the associated interest in protecting intellectual property is still present. In early 2022, the leader of the "Working Group for the Implementation of the Agenda 2030" stated, in a congress seminar, that the goals and objectives proposed in the agenda for 2030 are still far from being met and it is urgent that strategic knowledge of the parliamentary functions should be imbued with a sense of efficiency, effectiveness, and a framework of plurality. Additionally, in a conference titled, "Scientific evidence for parliamentary decision-making" the need for using scientific knowledge for law-related decisions was stated [10]. With this in mind, the scientific community must work on generating knowledge of all the benefits of amaranth and chia, including more evidence of the positive health effects of these grains' consumption in the diet and their active compounds, the agronomical advantages of their cultivation in comparison to other species, and the profit potential of the production chain of both grains, while achieving sustainable economic development, particularly in marginalized and indigenous populations.

4. Conclusions

Amaranth and chia have numerous attributes related to sustainability that make them an attractive crop to address Mexico's food self-sufficiency challenges. Additionally, their nutritional characteristics and positive health effects give these ancestral grains added value. In this context, their health benefits make them of interest to inventors, and the intellectual protection of the technologies derived from them has experienced an upturn in recent years. However, there is as yet no precise legal regulation of chia-derived products, and incentive policies for the cultivation of both grains have not been sufficient to encourage production growth, compared to other less profitable grains, such as maize. It is necessary to extend the scientific evidence of the benefits to society offered by these crops, including the social, economic, and ecological impact of both seeds, in order to encourage the creation of a legal framework that effectively stimulates the cultivation of both seeds in Mexico and contributes to the sustainable development of the country.

Author Contributions: Conceptualization, F.V.Z. and M.R.S.C.; methodology, F.V.Z. and M.R.S.C.; formal analysis, F.V.Z.; data curation, F.V.Z.; writing original draft, F.V.Z.; writing review and editing, F.V.Z. and M.R.S.C.; project administration, M.R.S.C.; funding acquisition, M.R.S.C. All authors have read and agreed to the published version of the manuscript.

Funding: This work was supported by grant Ia ValSe-Food-CYTED (119RT0567) and CONACYT scholarship 942394.

Institutional Review Board Statement: Not applicable.

Informed Consent Statement: Not applicable.

Data Availability Statement: Data available on request from the authors.

Conflicts of Interest: The authors declare no conflict of interest.

References

1. FAO. Alimentación y Agricultura Sostenibles [Internet]. Food and Agriculture Organization of the United Nations. Available online: http://www.fao.org/sustainability/es/ (accessed on 19 July 2022).
2. Valenzuela Zamudio, F.; Segura Campos, M.R. Amaranth, quinoa and chia bioactive peptides: A comprehensive review on three ancient grains and their potential role in management and prevention of Type 2 diabetes. *Crit. Rev. Food Sci. Nutr.* **2020**, *62*, 2707–2721. [CrossRef] [PubMed]
3. SIAP. SIACON [Software]. Available online: https://siga.impi.gob.mx/newSIGA/content/common/principal.jsfhttps://www.gob.mx/siap/documentos/siacon-ng-161430 (accessed on 1 July 2022).
4. IMPI. Búsqueda—SIGA [Internet]. Available online: https://siga.impi.gob.mx/newSIGA/content/common/principal.jsf (accessed on 1 July 2022).
5. NMX-FF-114-SCFI-2009. Diario Oficial de la Federación. Available online: https://dof.gob.mx/nota_detalle.php?codigo=5106745&fecha=25/08/2009#gsc.tab=0 (accessed on 19 July 2022).
6. NMX-FF-116-SCFI-2010. Available online: http://sitios1.dif.gob.mx/alimentacion/docs/NMX-FF-116-SCFI-2010_amaranto.pdf (accessed on 19 July 2022).
7. Mexican Law Initiative. Available online: http://sil.gobernacion.gob.mx/Archivos/Documentos/2021/04/asun_4179518_20210428_1619634670.pdf (accessed on 19 July 2022).
8. Mexican Law Reglas de Operación del Programa Producción para el Bienestar de la Secretaría de Agricultura y Desarrollo Rural para el Ejercicio Fiscal 2022. Available online: https://www.dof.gob.mx/nota_detalle.php?codigo=5646225&fecha=18/03/2022#gsc.tab=0 (accessed on 19 July 2022).
9. México sede del Primer Congreso Mundial del Amaranto [Internet]. Available online: https://masdemx.com/2018/10/amaranto-congreso-mundial-alimento-mexico/ (accessed on 19 July 2022).
10. Inauguran en San Lázaro el Curso "Evidencia Científica para la toma de Decisiones Parlamentarias" [Internet]. Available online: https://hojaderutadigital.mx/inauguran-en-san-lazaro-el-curso-evidencia-cientifica-para-la-toma-de-decisiones-parlamentarias/ (accessed on 19 July 2022).

Proceeding Paper

Nutrient Composition of Fresh Pasta Enriched with Chia (*Salvia hispanica* L.) †

Silvia Aja and Claudia Monika Haros *

Instituto de Agroquímica y tecnología de Alimentos (IATA-CSIC), 46980 Valencia, Spain
* Correspondence: cmharos@iata.csic.es

† Presented at the IV Conference Ia ValSe-Food CYTED and VII Symposium Chia-Link, La Plata and Jujuy, Argentina, 14–18 November 2022.

Abstract: Pasta is traditionally made from durum wheat semolina, but this can be substituted with other flour or semolina types and pasta can contain other ingredients. The nutritional characteristics depend on the ingredients, but, generally, pasta contains about 70% carbohydrate, mainly starch. Chia (*Salvia hispanica*) and their byoproducts are well-known for their high nutritional value, containing essential fatty acids (omega-3 and omega-6), a high mineral and vitamin content, and a high amount of fibre. The purpose of this study was to investigate the effects of different chia byproducts on pasta's technological parameters, their nutritional/functional characteristics (proximate composition, phytic acid) and a sensory evaluation (hedonic scale of nine points). The results showed a higher contribution of minerals in formulations with chia byproducts compared to the control sample. However, the mineral bioavailability could be compromised, as indicated by the phytic acid increment in formulations with chia byproducts. However, the glycaemic indexes were significantly similar to the control sample, with the exception of the samples with chia seeds. Regarding the technological characteristics of the formulations with chia, they did not show significant differences compared to the control sample. In this sense, the chia byproducts could be nutritional ingredients for use in pasta enrichment without depletion of the product quality.

Keywords: *Salvia hispanica* L.; chia byproducts; fresh pasta; nutritional characteristics

Citation: Aja, S.; Haros, C.M. Nutrient Composition of Fresh Pasta Enriched with Chia (*Salvia hispanica* L.). *Biol. Life Sci. Forum* **2022**, *17*, 3. https://doi.org/10.3390/blsf2022017003

Academic Editors: Norma Sammán, Mabel Tomás and Loreto Muñoz

Published: 19 October 2022

Publisher's Note: MDPI stays neutral with regard to jurisdictional claims in published maps and institutional affiliations.

1. Introduction

The presence of cereals in the diet has varied in terms of the evolution of the different patterns of food consumption. [1]. Pasta has been consumed in the Mediterranean countries for many centuries and takes second place, after bread, in world consumption [2]. For this reason, there is a food trend of enhancing pasta with quality food products to achieve nutritional improvements [3]. Dried pasta is traditionally made from durum wheat semolina, but this can be substituted with other flour or semolina types and pasta can contain other ingredients, e.g., eggs or spices [4]. The nutritional characteristics depend on the ingredients, but, generally, pasta contains about 70% carbohydrates, mainly starch [5].

In this sense, chia (*Salvia hispanica* L.) and their byproducts are well-known for their high nutritional value, as they contain essential fatty acids (omega-3 and omega-6) [4], have a high mineral and vitamin content, and are also an excellent source of dietary fibre [1]. This means that chia could be used to enrich food formulations.

The purpose of this study was to investigate the effects of different chia products (seeds, flour and by-products from cold-pressing oil extraction) on pasta's processing properties and to evaluate their nutritional value.

2. Materials and Methods

2.1. Raw Materials

Commercial Spanish wheat flour (W) was obtained from the local market. Chia seeds (CWS); chia whole flour (CWF), chia fibre (CF), and chia proteins (CP) were provided by BENEXIA Company (Santiago, Chile).

2.2. Pasta Production

The produced pasta was of the tagliatelle type, on a pilot scale, according to the specifications and procedure of the automatic pasta-maker (Nina, Springlane, Düsseldorf, Germany). Five types of pasta were produced: a control made of 100% wheat flour, and four fortified pastas with added chia seeds, chia whole flour, chia fibre and chia proteins (at a 10% substitute level).

2.3. Chemical Composition

Chemical analyses of the pasta were realized to determinate moisture (AOAC 925.09 Method) [6], protein (ISO/TS 16634-2) [7], total dietary fibre (TDF) and starch according to the approved AOAC 991.43 and 996.11, respectively [6].

The concentration of phytic acid (InsP_6) was determined as a phosphorus released by phytase and alkaline phosphatase by a quantity K-PHYT method, where the phosphate release was measured by a colorimetric technique (AOAC method 986.11) [8].

2.4. Preliminary Sensory Evaluation

The preliminary sensory evaluation was carried out by 20 untrained testers who consume pasta in their everyday life. The parameters that were evaluated were as follows: texture, appearance, colour, taste, odour and overall acceptability in a 9-point hedonic scale Iglesias-Puig et al. [9].

2.5. Statistical Analysis

Multiple sample comparison of the means (ANOVA) and Fisher's least significant differences (LSD) were applied to establish statistically significant differences between samples ($p < 0.05$).

3. Results

Regarding the results obtained for moisture and lipids, it was observed that the highest values were obtained in the control samples, followed by whole chia flour, chia fibre and chia proteins, respectively, as shown in Figure 1. In addition, the results obtained for protein and ash showed that the highest obtained value was for the chia protein sample, followed by chia fibre, whole chia flour and control sample. The value of starch on control sample was significantly higher than in samples with chia byproducts; however, the GI did not show significant differences between samples (Figure 2). The highest amount of phytic acid was in the formulation of chia fibre, followed by the whole chia flour, chia protein sample, and the lowest value was in control sample, as was expected (Figure 3).

Regarding the preliminary sensory evaluation, the highest values were obtained by the control paste with the maximal overall acceptability at values of around eight points, as was expected (Figure 4). A slight decline in the overall acceptability of pasta with chia byproducts was obtained, but close to the control, which good acceptability by consumers.

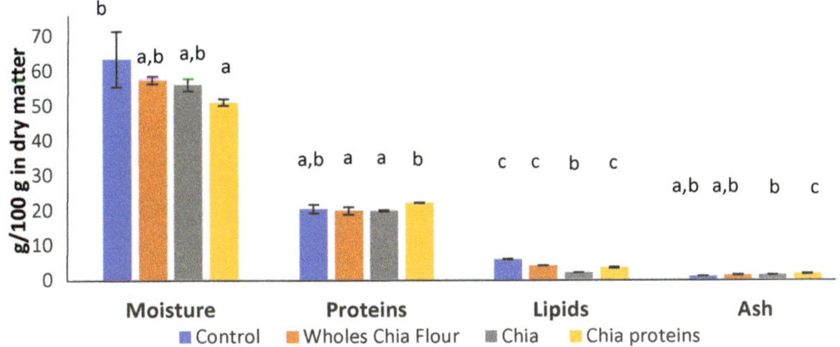

Figure 1. Fresh pasta proximate composition with chia byproducts. Values in bars of the same parameter followed by the same letter are not significantly different at 95% confidence level ($p < 0.05$).

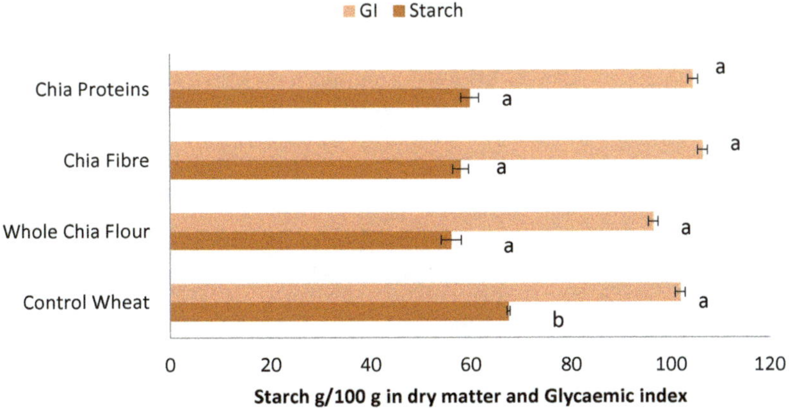

Figure 2. Starch concentrations and glycaemic index of paste with chia byproducts. Values in bars of the same colour followed by the same letter are not significantly different at 95% confidence level ($p < 0.05$).

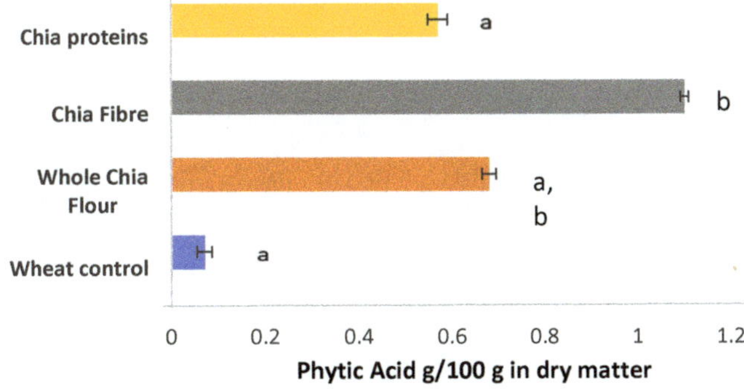

Figure 3. Acid phytic amount in pasta formulations with chia byproducts. Values in bars followed by the same letter are not significantly different at 95% confidence level ($p < 0.05$).

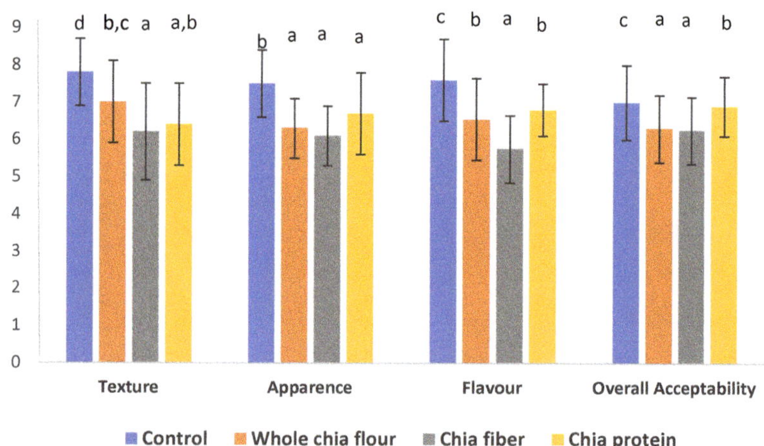

Figure 4. Preliminary sensory evaluation. Values in bars of the same colour followed by the same letter are not significantly different at 95% confidence level ($p < 0.05$).

4. Conclusions

Chia byproducts could be nutritional ingredients for use in the production of egg-free pasta enrichment without a depletion of product quality.

Author Contributions: Conceptualization, C.M.H.; methodology, S.A.; validation, S.A. and C.M.H.; formal analysis, S.A.; investigation, C.M.H.; resources, C.M.H.; writing—original draft preparation, S.A.; writing—review and editing, C.M.H.; visualization, S.A. and C.M.H.; supervision, C.M.H.; project administration, C.M.H.; funding acquisition, C.M.H. All authors have read and agreed to the published version of the manuscript.

Funding: This work was supported by grant Ia ValSe-Food-CYTED (119RT0567) and Food4ImNut (PID2019-107650RB-C21) from the Ministry of Sciences and Innovation (MICINN-Spain).

Institutional Review Board Statement: Not applicable.

Informed Consent Statement: Not applicable.

Data Availability Statement: Not applicable.

Conflicts of Interest: The authors declare no conflict of interest.

References

1. Tapsell, L.C.; Neale, E.P.; Satija, A.; Hu, F.B. Foods, Nutrients, and Dietary Patterns: Interconnections and Implications for Dietary Guidelines. *Adv. Nutr.* **2016**, *7*, 445–454. [CrossRef]
2. Webb, D. Pasta's History and Role in Healthful Diets. *Food Nutr.* **2019**, *54*, 213–220. [CrossRef]
3. Li, M.; Zhu, K.X.; Guo, X.N.; Brijs, K.; Zhou, H.M. Natural additives in wheat-based pasta and noodle products: Opportunities for enhanced nutritional and functional properties. *Compr. Rev. Food Sci. Food Saf.* **2014**, *13*, 347–357. [CrossRef] [PubMed]
4. Ullah, R.; Nadeem, M.; Khalique, A.; Imran, M.; Mehmood, S.; Javid, A.; Hussain, J. Nutritional and therapeutic perspectives of Chia (*Salvia hispanica* L.): A review. *J. Food Sci. Technol.* **2016**, *53*, 1750–1758. [CrossRef] [PubMed]
5. Biró, B.; Fodor, R.; Szedljak, I.; Pásztor-Huszár, K.; Gere, A. Buckwheat-pasta enriched with silkworm powder: Technological analysis and sensory evaluation. *LWT* **2019**, *116*, 108542. [CrossRef]
6. AOAC. Method 925.09, 991.43, 996.11, 986.11. In *Official Methods of Analysis*, 15th ed.; Association of Official Analytical Chemists: Arlington, VA, USA, 1996.
7. ISO/TS. *Food Products—Determination of the Total Nitrogen Content by Combustion According to the Dumas Principle and Calculation of the Crude Protein Content—Part 1 and 2: Cereals, Pulses and Milled Cereal Products (ISO/TS 16634-1 and ISO/TS 16634-2)*; International Organization for Standardization (ISO): Geneva, Switzerland, 2016.

8. McKie, V.; McCleary, B. A Novel and Rapid Colorimetric Method for Measuring Total Phosphorus and Phytic Acid in Foods and Animal Feeds. *J AOAC Int.* **2016**, *99*, 738–743. [CrossRef] [PubMed]
9. Iglesias-Puig, E.; Monedero, V.; Haros, M. Bread with Whole Quinoa Flour and Bifidobacterial Phytases Increases Dietary Mineral Intake and Bioavailability. *Food Sci. Technol.* **2015**, *60*, 71–77. [CrossRef]

Proceeding Paper

Preparation of Fresh Noodles with Chia and Amaranth †

Jannika Bailey, Silvia Farah, Pablo Mezzatesta and Emilia Raimondo *

Faculty of Nutrition Sciences, Juan Agustín Maza University, Mendoza M5519, Argentina
* Correspondence: emilia.raimondo@gmail.com
† Presented at the IV Conference Ia ValSe-Food CYTED and VII Symposium Chia-Link, La Plata and Jujuy, Argentina, 14–18 November 2022.

Abstract: Current nutritional recommendations lead to reformulate traditional products, such as fresh noodles, in order to improve the nutrients, they provide. The objective of this work was to determine the variation of the nutritional profile of fresh noodles by partially replacing wheat flour with chia and amaranth flour. For this purpose, wheat flour was partially substituted with 15% chia flour and 11% amaranth flour, respecting the proportions of the rest of the ingredients (called FS) and fresh noodles with unsubstituted wheat flour, which is taken as the standard (F). Once prepared, protein, total fat, ash, moisture, and fiber were determined by official analytical techniques, including carbohydrates by difference, energy value by calculation, and fatty acid profile by gas chromatography. From the analytical results, it appears that protein increased from 7.76 g% (F) to 10.87 g% (FS), carbohydrates decreased from 51.68 g% (F) to 39.87 g% (FS), and fiber increased from 3.19 g% (F) to 8.68 g% (FS). Total fats increased from 13.18 g% (F) to 18.68 g% (FS), of which omega-3 fatty acids increased from 0.67 g% (F) to 4.03 g% (FS). Energy value varies from 356 kcal/100 g (F) to 369 kcal/100 g (F). With the partial substitution of wheat flour with chia and amaranth, it was possible to improve the nutritional profile of the noodles, making them a feasible option for both industrial and home use.

Keywords: amaranth; chia; fresh noodles; lipidic profile; nutritional profile; vegetable protein

Citation: Bailey, J.; Farah, S.; Mezzatesta, P.; Raimondo, E. Preparation of Fresh Noodles with Chia and Amaranth. *Biol. Life Sci. Forum* **2022**, *17*, 4. https://doi.org/10.3390/blsf2022017004

Academic Editors: Norma Sammán, Mabel Cristina Tomás, Loreto Muñoz and Claudia Monika Haros

Published: 19 October 2022

1. Introduction

Due to the current regulations on the front of package warning seals, the food production industry needs to reformulate foods so that the products that reach the market do so with a better nutritional profile. On the other hand, the consumption of fresh pasta is frequent in Argentina, as well as in several countries of the region and in Europe [1].

According to Argentine legislation, the generic term "pasta or noodles" refers to non-fermented products obtained with mechanical mixing and kneading of baking flours or their mixtures with drinking water, with or without the addition of coloring substances authorized for this purpose, and with or without the addition of other permitted ingredients for this type of products. A mixture of mono- and diglycerides and monoglycerides of high concentration may be used in noodle products, alone or in a mixture, and in the technologically necessary quantity, without declaring it on the label [1].

To improve the nutritional profile of these pastas, ancestral seed flours such as amaranth (*Amaranthus caudatus*), which has a very good crop yield in Mendoza, Argentina, especially in the northwest of the country, and chia (*Salvia hispanica* L.) were chosen. Amaranth was chosen for its protein content of 16 to 18 g% [2], with a complete amino acid profile, resulting in a protein of high biological value [3]. "The Food and Agriculture Organization of the United Nations (FAO) recognizes this grain as a 'pseudocereal', with a higher protein content for human consumption, itis considered 'the food of the future' and recommended its intake in families, mainly in those with scarce resources" [4].

Chia seeds (*Salvia hispanica* L.) mainly provide omega-3 fatty acids (linolenic acid and its long chain derivatives), being the vegetable source with the highest content of this type of essential fatty acids, with the consequent benefit in human cardiovascular health. It also provides omega-6 fatty acids, its precursor being linoleic acid. The seeds are ground to obtain fine flour with an intense flavor, called "pinole" in Mexico, which is mixed with cereal flour to prepare different baked goods. This represents how these seed flours are used in this current work [5,6].

For all the above, the objective of this work was to determine the variation of the nutritional profile of fresh noodles by partially replacing wheat flour by chia and amaranth flour.

2. Materials and Methods

Fresh noodles were prepared in triplicate, partially substituting wheat flour 000 with 15% chia flour and 11% amaranth flour, respecting the proportions of the rest of the ingredients (this preparation was called FS) and fresh noodles with unsubstituted wheat flour, which was taken as the standard (called F).

2.1. Ways of Preparation

At the Applied Nutrition Research Laboratory of the Juan Agustín Maza University, Mendoza, Argentina, noodles were prepared on a pilot scale. For this purpose, two doughs were prepared in triplicate, one in which the 000 wheat flour was partially replaced by 15% chia flour and 11% amaranth flour, to which 17% fresh eggs were added so as not to add water, 10% high oleic sunflower oil, 0.3% salt, and 0.3% turmeric, which was called FS. And dough with 72% exclusively of 000 wheat flour, with the same proportion of the rest of the ingredients, this preparation was used as a standard, called F. A manual machine was used to perform the lamination and cutting of the dough. The centesimal composition analyses were performed in duplicate on the raw dough. In order to perform the sensory evaluation, the noodles were cooked in boiling water at 98 °C with Mendoza atmospheric pressure, in a ratio of one part noodles to three parts water for 3 minutes, starting from boiling. The noodles hydrate twice their weight; that is, from 100 g of noodles a 300 g dish is obtained. As such, it can be considered that the nutrients analyzed correspond to the ready to eat dish, taking into account that the fraction that leaches to the cooking water would correspond to the mineral fraction, not analysed in this study.

2.2. Laboratory Analysis

To determine the nutritional composition of both types of noodles (FS) and (F), the following methods were applied:

- *Humidity*: Method of AOAC 950.46 B. [7] Indirect method by drying in an oven at 100–105 °C, until constant weight is achieved.
- *Total fat*: Direct method by extraction with ethyl ether (crude fat), Soxhlet gravimetric method (A.O.A.C. 960.39, 1990) was used.
- *Cholesterol and Acid profile* by gas chromatography
- *Fibers*: Acid alkaline attack (AOAC, 15th edition 1990) was used.
- *Crude protein*: Kjeldahl method, (A.O.A.C. 928.08, 1990), determining nitrogen, using 6.25 as a protein conversion factor.
- *Ashes*: Direct Method (A.O.A.C. 923.03, 1990): by incineration in muffle (at 500 ± 10 °C), until constant ash weight.
- *Carbohydrates*: determined by difference, by the following formula:

$$100 - (\text{weight in grams [protein + fat + water + ash + fibers]}), \text{ in 100 g of food.}$$

- *Energy value*: by calculation

Energy value (kcal) = (protein *4) + (carbohydrates* 4) + (fat * 9). The conversion is 2000 kcal = 8400 kJ

2.3. Statistical Analysis

To analyze the assumption of normality, the Shapiro–Wilk test was applied. To analyze differences in means, the Student's *t*-test for independent samples was used. The analysis was carried out with the SPSS® statistical package.

2.4. Sensory Analysis

In order to evaluate the acceptability of the noodles, an acceptance test was carried out with 30 untrained judges.

3. Results

3.1. Noodles with and without Added Seed Flour

For all nutrients analyzed, the Shapiro–Wilk normality assumption $p = 0.526$ was met.

3.1.1. Protein

The addition of chia and amaranth flours increased the protein content from 7.76 ± 0.14 g% (F) to 10.87 ± 0.85 g% (FS). When applying the Student's t-test for independent samples, a statistically significant difference can be observed in protein $p = 0.003$.

3.1.2. Lipid Profile

Total fat increased from 13.18 ± 0.19 g% (F) to 18.66 ± 0.51 g% (FS). When applying the student's t-test, statistically significant differences p = 0.001 were found. If we analyze the different fractions that make up this lipid profile, we observe that saturated fats increase from 1.97 ± 0.12 g% (F) to 2.62 ± 0.23 g% (FS), with statistically significant differences between the two, p = 0.013. Monounsaturated fats practically do not vary from 9.14 ± 0.15 g% (F) to 9.62 ± 0.40 g% (FS), with no statistically significant differences between the two types of noodles. Cholesterol, provided by egg, is the same in both preparations 68.85 ± 0.73 mg% (F) and 68.85 ± 0.60 mg% (FS). Polyunsaturated fats increase from 2.07 ± 0.08 g% (F) to 6.44 ± 0.06 g% (FS), with statistically significant differences p = 0.001. If we analyze how these polyunsaturated fats are constituted, we observe that the contribution of omega 6 increases from 1.41 ± 0.80 g% (F) to 2.43 ± 0.05 g% (FS), with statistically significant differences p = 0.001. And the omega 3 intake increases from 0.67 ± 0.70 g% (F) to 4.02 ± 0.05 g% (FS), with statistically significant differences p = 0.001.

3.1.3. Carbohydrate Profile

Total carbohydrates decreased from 51.68 ± 0.99 g% (F) to 39.24 ± 0.23 g% (FS), with statistically significant differences $p = 0.001$. With a consequent increase in fiber content from 3.19 ± 0.04 g% (F) to 8.68 ± 0.39 g% (FS), with statistically significant differences $p = 0.001$.

3.1.4. Ash, Moisture, and Sodium Content

For these three nutrients there are no statistically significant differences when applying the Student's *t*-test for independent samples.

The ash content was 1.25 ± 0.04 g% (F) and 1.20 ± 0.03 g% (FS). Moisture was 22.94 ± 1.23 g% (F) and 21.35 ± 1.84 g% (FS).

The sodium content was 162 ± 6.25 g% (F) and 166 ± 5.29 g% (FS), which represents 7% of the daily sodium requirement.

3.1.5. Energy Value

The energy value of the noodles varied from 356 kcal/100 g for the common noodles to 369 kcal/100 g for the noodles with the addition of seeds. The increase in energy intake is not significant compared to the nutritional improvements obtained.

Table 1 shows all the values obtained for both noodles, allowing for ease of comparison.

Table 1. Composition of both noodles.

	Noodles (F)	Noodles with Seeds (FS)
Proteins g%	7.76 ± 0.14	10.87 ± 0.85
Total fats g%	13.18 ± 0.19	18.66 ± 0.51
Saturated fat g%	1.97 ± 0.12	2.62 ± 0.23
Monounsaturated fats g%	9.14 ± 0.15	9.62 ± 0.40
Polyunsaturated fats g% of which	2.07 ± 0.08	6.44 ± 0.06
Omega 6 g%	1.41 ± 0.80	2.43 ± 0.05
Omega 3 g%	0.67 ± 0.70	4.02 ± 0.05
Cholesterol g%	68.85 ± 0.73	68.85 ± 0.60
Carbohydrates g%	51.68 ± 0.99	39.24 ± 0.23
Dietary Fiber g%	3.19 ± 0.04	8.68 ± 0.39
Ash g%	1.25 ± 0.04	1.20 ± 0.03
Moisture g%	22.94 ± 1.23	21.35 ± 1.84
Sodium mg%	162 ± 6.25	166 ± 5.29
Energy Value kcal	356 ± 5	369 ± 8
Energetic Value kJ	1497 ± 22	1547 ± 34

4. Discussion

The addition of amaranth and chia flours improves the protein content. Taking into account that wheat has a protein of low biological value, the addition of amaranth improves the nutritional value of the noodles, making them ideal for low-income communities. In addition, fresh eggs provide protein of very high biological value.

The increase in total fat content is especially due to the lipid content of the chia seed, which is demonstrated by looking at the different lipid fractions. The almost constant contribution of monounsaturated fatty acids is due to the high oleic sunflower oil, which has a similar contribution to olive oil, but a more neutral taste, which facilitates the acceptability of the product. If the fatty acid profile is analyzed, noodles with seed flour are greatly improved because of the contribution of omega 9, which is beneficial to health, added to the increase in omega 3, which is extremely deficient in the population's diet, and necessary to improve the cardiovascular system, among other effects [7,8].

The decrease in total carbohydrate content, with a consequent increase in fiber, provided by seed flours, contributes to increased satiety [9].

Sodium content can be decreased by not adding salt at the time of preparation and replacing it with seasonings [10].

5. Conclusions

The addition of amaranth and chia seed flour to a noodle base dough improves the nutritional profile of the noodles and is a simple practice to apply in small-scale producers. The use of these additions increases the supply of healthier foods and revalue ancestral seeds, as well as being easy to use at home.

Author Contributions: Conceptualization, E.R.; methodology, S.F., J.B.; software, P.M.; validation, E.R. and P.M.; formal analysis, E.R.; statistical analysis, P.M.; investigation, all authors; resources, E.R.; data curation, P.M.; writing—original draft preparation, E.R.; writing—review and editing, S.F., J.B. and P.M.; visualization, S.F., J.B. and P.M.; supervision, E.R.; project administration, E.R.; funding acquisition, E.R. All authors have read and agreed to the published version of the manuscript.

Funding: This work was funded by the Juan Agustin Maza University and by La ValSe-Food-CYTED (119RT0567).

Institutional Review Board Statement: Not applicable.

Informed Consent Statement: Not applicable.

Data Availability Statement: The works consulted are detailed in the bibliography.

Conflicts of Interest: The authors declare no conflict of interest.

References

1. Law 27.642 "Promotion of Healthy Eating". 2022. Available online: https://www.boletinoficial.gob.ar/detalleAviso/primera/252728/20211112 (accessed on 13 June 2022).
2. Suárez, P.A.; Martínez, J.G.; Hernández, J.R. Amaranto: Efectos en la Nutrición y la Salud. *Tlatemoani: Revista Académica de Investigación Número 12, abril 2013*. Available online: https://dialnet.unirioja.es/servlet/articulo?codigo=7324335 (accessed on 1 June 2022).
3. Gonor, K.V.; Pogozheva, A.V.; Derbeneva, S.A.; Trushina, E.N.; Mustafina, O.K. The influence of a diet with including amaranth oil on antioxidant and immune status in patients with ischemic heart disease and hyperlipoproteidemia. *Vopr. Pitan.* **2006**, *75*, 30–33. [PubMed]
4. FAO. Agronoticias: Actualidad Agropecuaria de América Latina y el Caribe. 2014. Available online: https://www.fao.org/in-action/agronoticias/detail/es/c/514662/ (accessed on 6 June 2022).
5. Zavalía, L.; Alcocer, M.G.; Fuentes, F.J.; Rodriguez, W.A.; Morandini, M.; Devani, M.R. Desarrollo del cultivo de Chía en Tucumán, República Argentina. *Avance Agroind.* **2011**, *32*, 27–30.
6. Xingú López, A.; González Huerta, A.; de La Cruz Torres, E.; Sangerman Jarquín, D.M.; Orozco de Rosas, G.; Rubí Arriaga, M. Chía (*Salvia hispanica* L.) situación actual y tendencias futuras. Revista Mexicana de Ciencias Agrícolas. *Rev. Mex. Cienc. Agrícolas* **2017**, *8*, 1619–1631. [CrossRef]
7. Carrillo-Gómez, C.S.; Gutiérrez-Cuevas, M.; Muro-Valverde, M.; Martínez-Horner, R.; Torres-Bugarin, O. La chía como super alimento y sus beneficios en la salud de la piel. *El Resid.* **2017**, *12*, 18–24.
8. Travieso, J.C.F. Ácidos Grasos Omega-3 y Prevención Cardiovascular. *Rev. CENIC Cienc. Biológicas* **2010**, *41*, 3–15. Available online: https://www.redalyc.org/pdf/1812/181221644001.pdf (accessed on 1 June 2022).
9. Pino Villalón, J.L.; Rojas Muñoz, M.; Orellana Saez, B.; Torres Mejías, J. Fortificación con fibra dietética como estrategia para aumentar la saciedad: ensayo aleatorizado doble ciego controlado. *Rev. Esp. Nutr. Hum. Diet.* **2020**, *24*, 336–344. [CrossRef]
10. Gaitán, D.; Chamorro, R.; Cediel, G.; Lozano, G.; Gomes, F.D.S. Sodio y Enfermedad Cardiovascular: Contexto en Latinoamérica. *ALAN* **2015**, *65*, 206–215.

Proceeding Paper

Utilization of Hydrothermally Treated Flours in Gluten-Free Doughs †

Maria A. Gimenez *, Cristina N. Segundo, Natalia E. Dominguez, Norma C. Samman and Manuel O. Lobo

CIITED Centro de Investigación Interdisciplinario en Tecnología y Desarrollo Social del NOA—CONICET, Facultad de Ingeniería, Universidad Nacional de Jujuy, Ítalo Palanca 10, San Salvador de 4600, Argentina
* Correspondence: malejandragimenez@gmail.com

† Presented at the IV Conference Ia ValSe-Food CYTED and VII Symposium Chia-Link, La Plata and Jujuy, Argentina, 14–18 November 2022.

Abstract: Hydrothermal treatments are suitable for modifying the physicochemical properties of flours because they favor the total or partial gelatinization of starch, allowing products with different rheological and water absorption capacities to be obtained. The objective of the study was to apply the processes of extrusion cooking (CE), alkaline extrusion (OHE) and traditional nixtamalization (N) in four races of native maize from the province of Jujuy, Argentina (perlita, cuzco, chulpi and culli) to obtain flours with adequate aptitudes for use in gluten-free doughs. The different processes and the characteristics of each race had a significant effect ($p < 0.05$) on the hydration properties of the flours, both factors showing a greater effect on flour from the chulpi race, indicated by the increase in its hydration properties. Therefore, in the textural properties, the elasticity and the viscosity indices of the dough were dependent on the races and the processes, influencing these properties by the subjective water capacity. CE gives greater elasticity to the dough, presenting the highest values for the dough of the perlita race (5.58 mm). OHE provided a lower viscosity index (0.03 N $\times$ s) in the dough of the perlita race, indicating a poor integration of the flour components. The N did not confer remarkable properties to the dough, showing breakage of the dough touched. It is important to note that the OHE process provided dough with adequate properties due to intermediate values of elasticity (4.47–4.5 mm) and resistance to kneading (0.34–0.078 N $\times$ s). This study will optimize the development of extensible dough from the chulpi and culli races.

Keywords: alkaline extrusion; dough; gluten-free; nixtamalization; maize

Citation: Gimenez, M.A.; Segundo, C.N.; Dominguez, N.E.; Samman, N.C.; Lobo, M.O. Utilization of Hydrothermally Treated Flours in Gluten-Free Doughs. *Biol. Life Sci. Forum* **2022**, *17*, 5. https://doi.org/10.3390/blsf2022017005

Academic Editors: Mabel Cristina Tomás, Loreto Muñoz and Claudia Monika Haros

Published: 20 October 2022

1. Introduction

The native maize from the province of Jujuy, Argentina, represent the base of pre-Columbian food culture, especially in the regions of Puna, Quebrada and Valles [1]. The different races of corn have different technofunctional properties, indicating that the application of different processes of the food industry [2] would allow to obtain flours with properties adequate for use in various food products. Hydrothermal treatments are very attractive to modify the functional properties of cereal flours. Traditional nixtamalization is a process that consists of cooking corn kernels in a water–alkaline agent solution. This process converts the hemicellulose of the cell wall into soluble gums, gelatinizes the starch, saponifies part of the lipids, releases niacin and solubilizes part of the proteins that surround the starch granules, obtaining extensible masses to make taco tops [3]. In extrusion cooking, the flours are subjected to high temperatures and mechanical shearing at relatively low levels of moisture content that allows the pregelatinization of the starch, the denaturation of the protein, which allows changing the rheological properties of the flours [4]. Alkaline extrusion is an environmentally friendly technological alternative that is similar to traditional nixtamalization [5]. These processes directly affect the physicochemical properties of the cereal starch and the viscoelastic properties of the flour.

The objective of the study was to apply the processes of extrusion cooking (EC), alkaline extrusion (EOH) and traditional nixtamalization (N) in four races of native maize from the province of Jujuy, Argentina (perlita, cuzco, chulpi and culli) to obtain flours with suitable aptitudes for use in gluten-free doughs.

2. Materials and Methods

2.1. Raw Materials

Four races of native maize of different endosperm from the province of Jujuy, Argentina: perlita, cuzco, chulpi and culli were acquired from local producers in Ocumazo, Humahuaca, Jujuy, Argentina. Some of the grains of each race were integrally ground in a hammer mill to a particle size of less than 420 µm, and some were left unground.

2.2. Hydrothermal Processes

Cooking extrusion (CE)

Whole flour was conditioned at 28% of moisture 2 h before the extrusion cooking process. A Brabender KE 19 single-screw extruder with 2:1 compression ratio and 3 mm nozzle was used. The process was carried out at a screw speed of 60 rpm and a feed speed of 20 rpm. The extruder equipment managed three heating zones: 40 °C (feeding zone) −60 °C (compression zone) and 80 °C (cooking zone).

Alkaline extrusion (OHE)

At 12 hours before the process, 0.25 g of Ca $(OH)_2$ /100 g of flour was added to each whole meal flour sample, and then they were conditioned at 28% humidity. Each sample was mixed for 3 min and stored in a polyethylene bag in the refrigerator. The OHE was carried out in the same conditions as CE. The extrudates obtained by these two processes were collected in a tray and dried in an oven at 30 °C for 12 h.

Nixtamalization (N)

The corn grains of the different races were nixtamalized following the procedure of Bello-Perez et al. [2], with some modifications. The grains were mixed with water in a 1:3 ratio, and 2% Ca $(OH)_2$ was added. The cooking was carried out for 40 minutes and then left to rest for 24 hours. After nixtamalization, the liquid was drained, and the nixtamalized maize was washed with running tap water for 5 minutes until calcium hydroxide was eliminated and then dried at 40 °C in a forced convection oven. After applying these three processes, the samples obtained were ground in a hammer mill to a particle size of ≤250 µm.

2.3. Hydration Properties of Hydrothermally Treated Flours

Water absorption (WAI) and water solubility (WSI) indices

To determine the water absorption (WAI) and the water solubility (WSI) indices of the flours treated, the methodology described by Sing et al. [6] was followed. The experiment was carried out in triplicate. WAI and WSI were calculated by using the following expressions.

Subjective Water Absorption Capacity (SWAC)

SWAC was determinate according to Gaitan-Martinez et al [7].

2.4. Textural Properties of Dough

A total of 200 g of the flours obtained by extrusion cooking (CE), alkaline extrusion (EOH) and nixtamalization (N) were rehydrated according to the subjective water absorption capacity (SWAC) to obtain the respective masses. The fresh dough was divided into discs of 25 grams, 8 cm in diameter and 1 cm thick. All masses were analyzed after 12 h of refrigeration. The discs obtained from fresh masses were subjected to a puncture test of 40% of the original height with a flat-tip SMSP/3 probe (3 mm diameter) and a 25 kg cell. The test speed was 1 mm/s, and six disks were tested for each process and race. The following parameters were analyzed: penetration force in Newton (N) and elasticity (mm) and viscosity indices, also called cohesion work expressed in units of N × s.

2.5. Statistical Analysis

Multiple analyses of variance were used to determine the individual effects of thermal treatment and the corn race. Fisher's least significant differences test was used to calculate the means with 95% confidence intervals. The statistical analysis was performed with the XLStat trial (Copyright © 2015 Addinsoft, Paris, France).

3. Results and Discussion

3.1. Gel Hydration Properties (WAI and WSI) and Subjective Water Absorption Capacity (SWAC)

Gel hydration properties are important in bulking and depend on proteins and polysaccharides.

The race factor had a statistically significant effect on WSI ($p < 0.05$), mainly in the meal of the chulpi race, indicated by the increase in WSI in all cases (Table 1). This can be attributed to the higher protein content (8–13%) of chulpi and the higher protein compaction in its vitreous endosperm [8] compared to the protein content of the other races (7–11%) reported by Gimenez et al. [2] generating less interaction between proteins and water, increasing dissolved solids. The simple ANOVA analysis showed that the different processes generated significant differences in the WSI and WAI of the flours. Showing in most cases that CE and N caused the lowest WSI value and the highest WAI value, but not for the culli race, which did not show significant differences in the gel hydration properties with respect to the processes. In CE and N, the content of soluble fibers and complexes that involve starch could have increased, which favored the formation of a compact structure that decreased accessibility to water, obtaining low WSI values [4]. On the other hand, it is possible that the presence of a high percentage of damaged starch has an impact on a higher IAA. The Subjective Water Absorption Capacity (SWAC) is an important property from an economic point of view because it impacts the yield of the flour to dough conversion. The multifactorial analysis indicated that only the race factor had a statistically significant effect on the SWAC (Table 1). These values ranged from 0.74 to 0.96 ml water/g flour. It was observed that with the cuzco race a greater amount of water was needed to form dough and less was needed with the chulpi race, for all the processes. With OHE, the process was generally observed with the highest values. This property influences the texture parameters of the dough since it is related to the interaction of the components of the flour and its ability to bind water [7].

Table 1. Gel hydration properties of maize flour from the extrusion cooking (CE), alkaline extrusion (EOH) and nixtamalization (N) processes.z.

Race	Process	WAI (%)	WSI (*) (g/g)	SWAC (mL/g)
Cuzco	CE	3.80^{b} **	0.05^{b} **	0.93^{b} **
	OHE	4.46^{c} **	0.06^{c} **	0.96^{b} **
	N	2.65^{a} **	0.03^{a} **	0.84^{a} **
Chulpi	CE	3.16^{b} **	0.115^{a} **	0.74^{a} **
	OHE	2.62^{a} **	0.18^{b} **	0.84^{b} **
	N	3.61^{c} **	0.205^{c} **	0.93^{c} **
Perlita	CE	3.68^{b} **	0.065^{a} **	0.89^{c} **
	OHC	3.24^{a} **	0.08^{c} **	0.86^{b} **
	N	3.09^{a} **	0.05^{b} **	0.81^{a} **
Culli	CE	4.15^{b} **	0.055^{a} **	0.85^{b} **
	OHC	3.57^{ab} **	0.055^{a} **	0.90^{c} **
	N	3.21^{a} **	0.045^{a} **	0.81^{a} **

WAI: water absorption index. WSI: water solubility index. SWAC: subjective water absorption capacity. * Indicates that the race factor had a statistically significant effect on WSI (>0.05). ** Indicates the significant differences by process by race. Values followed by different letters within each parameter.

3.2. Textural Properties of Dough

Figure 1a–c show the firmness, elasticity and viscosity indices of the dough of the flours of different races treated with EC, OHE and N. In the puncture test, the firmness of the dough was measured as the maximum force recorded during penetration (Figure 1a). The analysis of variance showed that the applied processes did not have a significant impact ($p < 0.05$) on this property. Regarding this property, the chulpi dough showed a different behavior from that of the other races. The chulpi dough with nixtamalized corn flour presented the greatest firmness; this may be due to the particular characteristics of the endosperm of this race characterized by the presence of crystallized sugars [8]. Alkaline hydrothermal treatments also solubilize the fibers of the pericarp, allowing greater interaction with the water, contributing to the formation of firmer dough. In the dough of the other races, the shear forces in the EC process could have produced structural changes in the macrocomponents of the flours that favored the interaction with the water, increasing their firmness. The viscosity index (Figure 1b) showed a significant dependence ($p < 0.05$) with the applied treatments. This parameter is related to the cohesion of the dough and indicates the resistance to tearing during kneading. The OHE process presented a higher value in the viscosity index (0.44 N $\times$ s) for the cuzco breed, indicating a greater integration of the components that are part of the dough matrix, while the perlite flour mass presented the lowest value. The Figure 1c shows the elasticity presented by the masses of the treated flours. In this parameter, both the process applied and the race of the maize had a significant effect. The EC process provides greater elasticity in the masses, observing the highest value in the mass of perlite (5.58 mm). The addition of the alkaline agent to the process had a significant decreasing effect ($p < 0.05$), finding the lowest value in the dough of the nixtamalized culli race (3.02 mm). High elasticity values are undesirable in laminated dough; however, dough with low elasticity values breaks when it is laminated [9].

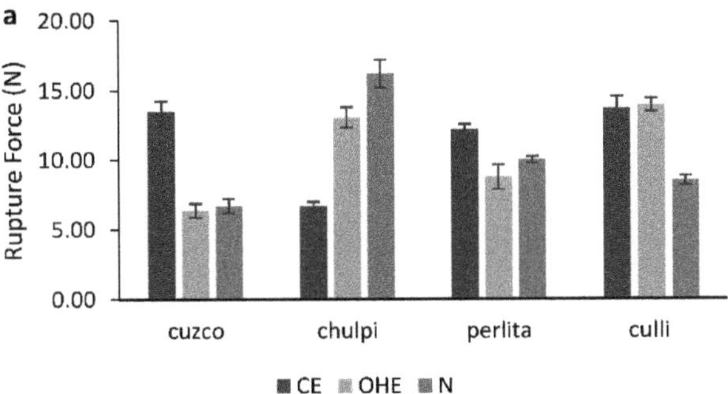

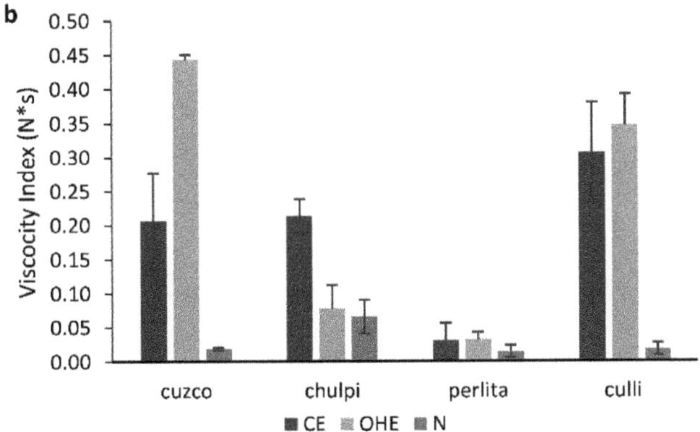

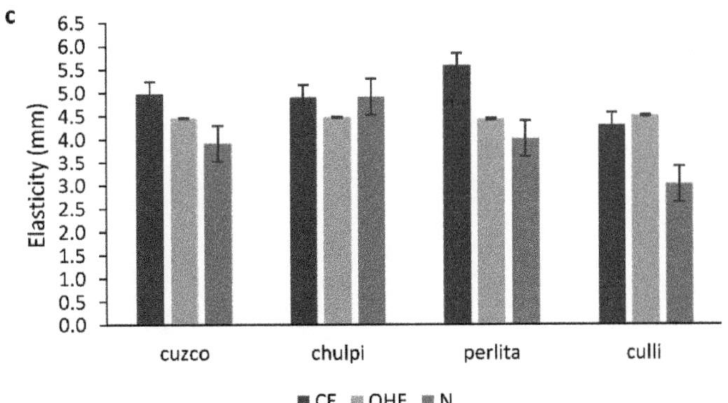

Figure 1. (**a**) Rupture Force (N); (**b**) Viscosity Index (N × s); (**c**) Elasticity (mm) of dough of the races cusco, chulpi, perlita and culli from the CE, OHE, N processes.

4. Conclusions

The processes applied and the characteristics of each race had an important effect on the textural properties of their masses. In particular, the characteristics of the endosperm of the chulpi race give it properties and behavior that were different from the rest of the races studied. The alkaline extrusion process provided dough with intermediate values of elasticity, firmness and good resistance to kneading. The chulpi breed presented the best aptitude to be used in gluten-free dough.

Author Contributions: Conceptualization and methodology M.A.G. and C.N.S.; validation, M.A.G., M.O.L. and N.C.S.; formal analysis, N.E.D.; investigation, N.E.D.; resources, M.O.L. and N.C.S.; data curation, C.N.S.; writing—original draft preparation, N.E.D. and C.N.S.; writing—review and editing, M.A.G.; visualization, C.N.S.; supervision, M.A.G.; project administration M.O.L. and M.A.G.; funding acquisition, M.A.G. and N.C.S. All authors have read and agreed to the published version of the manuscript.

Funding: This work was supported by a grant from Ia ValSe-Food-CYTED (119RT0567), CONICET and SECTER, National University of Jujuy (Argentina).

Institutional Review Board Statement: Not applicable.

Informed Consent Statement: Not applicable.

Data Availability Statement: Not applicable.

Conflicts of Interest: The authors declare no conflict of interest.

References

1. Ramos, R.S.; Hilgert, N.I.; Lambaré, D.A. *Agricultura Tradicional y Riqueza de Maíces (Zea mays);* Estudio de Caso en Caspalá: Provincia de Jujuy, Argentina, 2013. (In Spanish)
2. Giménez, M.A.; Segundo, C.N.; Lobo, M.O.; Sammán, N.C. Physicochemical and Techno-Functional Characterization of Native Corn Reintroduced in the Andean Zone of Jujuy, Argentina. *Proceedings* **2020**, *53*, 7. [CrossRef]
3. Bello Pérez, A.B.; Díaz, P.O.; Acevedo, E.A.; Santiago, C.N.; López, O.P. Propiedades químicas, fisicoquímicas y reológicas de masas y harinas de maíz nixtamalizado. *Agrociencia* **2002**, *36*, 319–328. (In Spanish)
4. Martínez, M.M.; Pico, J.; Gómez, M. Physicochemical modification of native and extruded wheat flours by enzymatic amylolysis. *Food Chem.* **2015**, *167*, 447–453. [CrossRef] [PubMed]
5. Enríquez-Castro, C.M.; Torres-Chávez, P.I.; Ramírez-Wong, B.; Quintero-Ramos, A.; Ledesma-Osuna, A.I.; López-Cervantes, J.; Gerardo-Rodríguez, J.E. Physicochemical, rheological, and morphological characteristics of products from traditional and extrusion nixtamalization processes and their relation to starch. *Int. J. Food Sci.* **2020**, *2020*, 5927670. [CrossRef] [PubMed]
6. Singh, N.; Singh, J.; Kaur, L.; Sodhi, N.S.; Gill, B.S. Morphological, thermal and rheological properties of starches from different botanical sources. *Food Chem.* **2003**, *81*, 219–231. [CrossRef]
7. Gaytán-Martínez, M.; Figueroa, J.D.C.; Vázquez-Landaverde, P.; Morales-Sánchez, E.; Martínez-Flores, H.E.; Reyes-Vega, M.L. Caracterización fisicoquímica, funcional y química de harinas nixtamalizadas de maíz obtenidas por calentamiento óhmico y proceso tradicional. *CyTA-J. Food* **2012**, *10*, 182–195. [CrossRef]
8. Salvador-Reyes, R.; Rebellato, A.P.; Pallone, J.A.L.; Ferrari, R.A.; Clerici, M.T.P.S. Kernel characterization and starch morphology in five varieties of Peruvian Andean maize. *Food Res. Int.* **2021**, *140*, 110044. [CrossRef] [PubMed]
9. Topete-Betancourt, A.; Santiago-Ramos, D.; Figueroa-Cárdenas, J.D.D. Relaxation tests and textural properties of nixtamalized corn masa and their relationships with tortilla texture. *Food Biosci.* **2020**, *33*, 100500. [CrossRef]

Proceeding Paper

Nutritional Characterization of Ancestral Organic Wheats: Emmer, Khorasan and Spelt †

Aikaterini Athinaiou [1,2], Silvia Aja [1] and Claudia Monika Haros [1,*]

[1] Instituto de Agroquímica y Tecnología de Alimentos (IATA), Consejo Superior de Investigaciones Científicas (CSIC), Av. Agustín Escardino 7, Parque Científico, Paterna, 46980 Valencia, Spain
[2] Department of Food Science and Human Nutrition, Agricultural University of Athens (AUA), Av. Iera Odos 75, 11855 Athina, Greece
* Correspondence: cmharos@iata.csic.es
† Presented at the IV Conference Ia ValSe-Food CYTED and VII Symposium Chia-Link, La Plata and Jujuy, Argentina, 14–18 November 2022.

Abstract: Nowadays, consumers show a growing interest in the consumption of foods made with ancestral grains, the main components of the diet of our ancestors. The ancestral grains come from millenary cultivars and have now burst onto the international market as part of a nutritious and healthy diet. Some of these crops refer to ancestral wheats. The objective of this study was to determine the nutritional characteristics of ancient wheats compared to the modern one. Ancient crops such as emmer (*Triticum dicoccum*, known as "farro medio" or "farro"), khorasan (*Triticum turanicum*, the best known kamut) and/or spelt (*Triticum aestivum* ssp. spelta, known as "escanda" or "farro grande") were the raw materials of the current investigation. Characterization of wheat seeds/whole flours in terms of moisture, ash, total dietary fiber, proteins, and lipids, phytates and phytase activity were determined. In general, these analyses do not support the suggestion that ancient wheats are generally more nutritious and/or healthy than modern wheats. The results support the consumption recommendation of the intake of whole grains (modern or ancients) to prevent non-transmissible illnesses.

Keywords: ancestral grains; emmer; khorasan; spelt; nutritional characteristics

Citation: Athinaiou, A.; Aja, S.; Haros, C.M. Nutritional Characterization of Ancestral Organic Wheats: Emmer, Khorasan and Spelt. *Biol. Life Sci. Forum* **2022**, *17*, 6. https://doi.org/10.3390/blsf2022017006

Academic Editor: Norma Sammán, Mabel Cristina Tomás and Loreto Muñoz

Published: 23 October 2022

1. Introduction

Wheat is one of the oldest crops. The origin and evolution of wheat has been the subject of study over the years [1]. The development of new special foods based on grain blends has permitted the use of the so called "ancient wheats" as components that convey naturalness, unconventional and nutritional properties [2]. The attention towards these ancient species have also been renewed by the increasing demand for traditional products [2]. However, about 95% of the wheat produced is Triticum aestivum, a hexaploid species usually called "common", "bread" or "soft" wheat [3] and the ancestors represent the remainder of the percentage.

The ancient wheats khorasan (*Triticum turgidum* ssp., the best known kamut, tetraploid), emmer (*Triticum dicoccum*, knowing as "farro medio" or "farro", tetraploid) and spelt (*Triticum aestivum* ssp. spelta, knowing as "escanda" or "farro grande", hexaploid) have been cultivated in very low amounts compared to the common wheat species (*T. aestivum* L., hexaploid) [4].

On the other hand, a meta-analysis confirmed the association between the consumption of whole grains and a substantial and significant decreased risk for cardiovascular disease, cancer which reveals the importance of consuming whole grains [5].

The purpose of this study was to determine the nutritional characteristics of ancient wheats compared to the modern ones.

2. Materials and Methods

2.1. Materials

Ancient crops such as khorasan, spelt organic and commercial common wheat were found at local Spanish market, whereas emmer was found at a Greek local market. Figure 1 shows the ancient wheats used in this study.

Figure 1. Organic ancestral wheats. (**a**) Emmer; (**b**) Spelt; (**c**) Khorasan.

2.2. Composition of the Raw Materials

Proximate analyses of the raw materials were performed in terms of moisture (AOAC 925.09 Method) [6], protein (ISO/TS 16634-2) [7], lipid (AACC Method 30-10) [8], ash contents (AACC Method 08-03) [8] and total dietary fibre (TDF) (AOAC Method 991.43) [6]. Measurements were taken three times.

Determination of *Myo-Inositol* Hexakiphosphate. The concentration of *myo*-inositol hexakisphosphate or phytic acid (InsP_6) was determined as phosphorus released by phytase and alkaline phosphatase by a quantity K-PHYT method [9]. The phosphate released was measured by a colorimetric technique (AOAC Method 986.11) [6]. All measurements were taken three times.

Phytase activity. The phytase activity was determined according to the methodology described by Haros et al. [10] expressed in U/g of wheat, whereas one unit (U) was defined as 1.0 µmol of Pi liberated per hour at 50 °C and pH: 5.5.

Statistical Analysis. To establish significant differences between samples one-way ANOVA and Fisher's least significant differences (LSD) were applied, all differences were considered significant at $p < 0.05$.

3. Results

The moisture values obtained for the wheats analyzed were very similar, the high moisture value is for spelt, control and organic wheat, emmer, followed by khorasan, respectively. It should be noted that khorasan wheat has the highest protein value of the ancestral varieties studied (approximately more than 30% respect to the control) followed by spelt wheat (Figure 2). The other wheats, such as emmer, control and organic, presented significantly lower values than the previous ones. It is important to remark that all wheats have the proteins which form the gluten. However, ancient wheats, despite having higher protein contents, they do not present higher gluten (data does not show).

Regarding the lipid values, we observe that the highest value obtained corresponds to spelt, followed by khorasan wheat, emmer, organic and control with values very similar between them. It is observed that the total dietary fiber values were very similar, with higher values for organic wheat, followed by khorasan, spelt and control wheat, and the lowest value for emmer wheat (Figure 2).

The highest values for ash were obtained for spelt and emmer wheats, followed by organic and khorasan and very closely continued by the control wheat. In terms of phytic acid concentration the highest value corresponds to spelt, followed by organic and khorasan wheat, because the values were very similar, and the next lowest was the control and emmer wheat (Figure 3). The phytic acid forms the phytate salts with minerals, which are the mineral storage in plants. On the other hand, phytase is the enzyme which hydrolyses the phytic acid/phytate in lower *myo*-isnositol phosphate, realizing inorganic phosphate. The phytase activity in all the investigated whole wheat flours presented significantly similar values between samples and according to previous works (Figure 4).

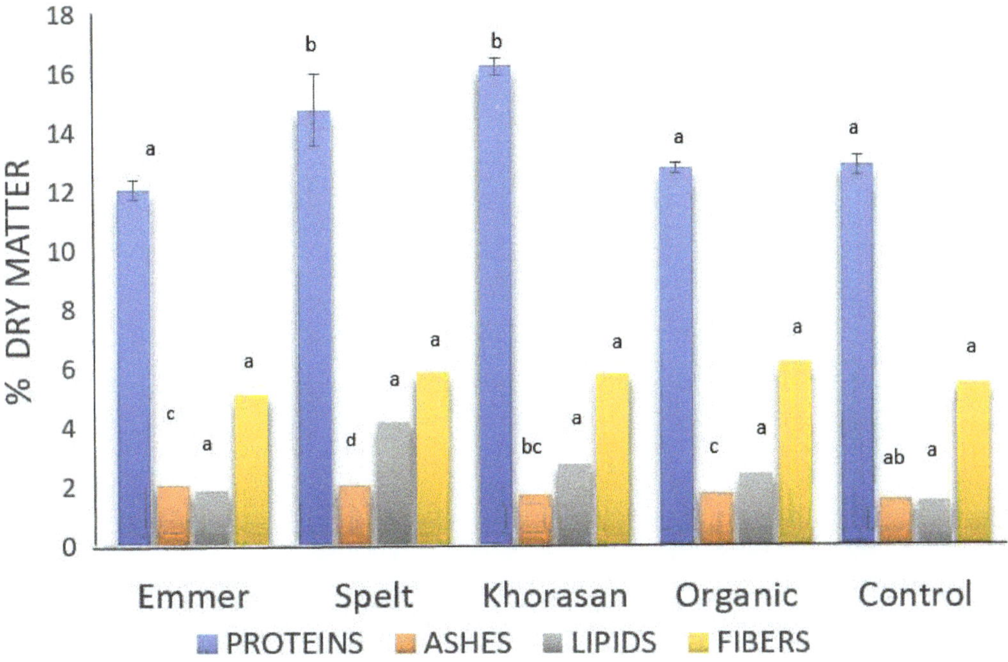

Figure 2. Proximate composition of the organic ancestral wheats comparing to modern wheats (organic and control). Data are expressed as mean ± standard deviation ($n = 3$), Values in bars followed by the same letter, in the same parameter, are not significantly different at 95% confidence level ($p < 0.05$).

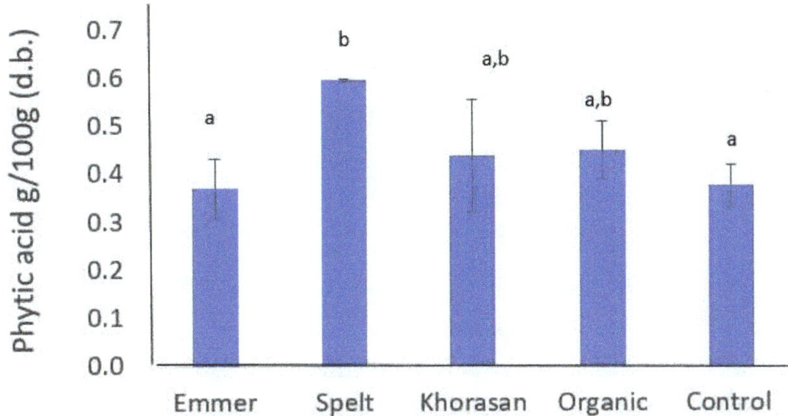

Figure 3. Phytic Acid content in ancestral wheats. Data are expressed as mean ± standard deviation ($n = 3$), Values in bars followed by the same letter are not significantly different at 95% confidence level ($p < 0.05$). Dry basis, d.b.

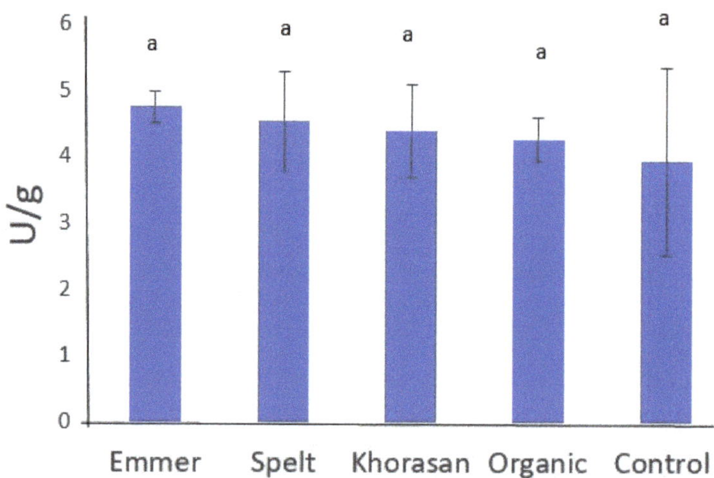

Figure 4. Phytase Activity. Data are expressed as mean ± standard deviation ($n = 3$), Values in bars followed by the same letter are not significantly different at 95% confidence level ($p < 0.05$).

4. Conclusions

Protein content was referred to as the most important factor that affects bread-making and baking quality. In our study khorasan wheat represents the highest value, but this value does not mean that they correspond to a significant amount of gliadins and gutenins, which are the proteins that constitute gluten. According to the ash values, the results for the seed spelt and emmer confirms their high nutritional value of the seeds compared to the other varieties. However, phytic acid has been considered as an anti-nutrient factor in human nutrition due to its effect on the inhibition of mineral bioavailability. According to our results, the spelt sample presented the highest percentage of mineral but also the highest phytate content, which could affect their bioavailability. The phytase activity was similar in all the studied wheats, but there was a tendency for higher values in ancestral wheats compared to the control wheats, which could produce the phytic acid degradation more efficiently during the fermentation step in the development of cereal products. Although the results cannot confirm that ancestral wheats are more nutritious than common wheat. The whole grain intake is still being the best strategy to obtain the high amount of nutrients and bioactive of cereal grains.

Author Contributions: Conceptualization, C.M.H.; methodology, A.A. and S.A.; validation, S.A. and C.M.H.; formal analysis, A.A. and S.A.; investigation, A.A. and C.M.H.; resources, C.M.H.; writing—original draft preparation A.A. and S.A.; writing—review and editing, C.M.H.; visualization, A.A. and S.A.; supervision, C.M.H.; project administration, C.M.H.; funding acquisition, C.M.H. All authors have read and agreed to the published version of the manuscript.

Funding: This work was supported by grant Ia ValSe-Food-CYTED (119RT0567) and Food4ImNut (PID2019-107650RB-C21) from the Ministry of Sciences and Innovation (MICINN-Spain).

Institutional Review Board Statement: Not applicable.

Informed Consent Statement: Not applicable.

Data Availability Statement: Not applicable.

Acknowledgments: The fellowship of Aikaterini Athineou from the ERASMUS+ Program EU is gratefully acknowledged.

Conflicts of Interest: The authors declare no conflict of interest.

References

1. Dvorak, J.; Deal, K.R.; Luo, M.-C.; You, F.M.; von Borstel, K.; Dehghani, H. The Origin of Spelt and Free-Threshing Hexaploid Wheat. *J. Hered.* **2012**, *103*, 426–441. [CrossRef] [PubMed]
2. Lacko-Bartošová, M.; Čurná, V. Nutritional characteristics of emmer wheat varieties. *J. Microbiol. Biotechnol. Food Sci.* **2015**, *4*, 95–98. [CrossRef]
3. Mayer, K.F.X. A chromosome-based draft sequence of the hexaploid bread wheat (*Triticum aestivum*) genome. *Science* **2014**, *345*, 286.
4. Geisslitz, S.; Longin, C.F.H.; Scherf, K.A.; Koehler, P. Comparative Study on Gluten Protein Composition of Ancient (Einkorn, Emmer and Spelt) and Modern Wheat Species (Durum and Common Wheat). *Foods* **2019**, *8*, 409. [CrossRef] [PubMed]
5. Aune, D.; Keum, N.; Giovannucci, E.; Fadnes, L.T.; Boffetta, P.; Greenwood, D.C.; Tonstad, S.; Vatten, L.J.; Riboli, E.; Norat, T. Whole grain consumption and risk of cardiovascular disease, cancer, and all cause and cause specific mortality: Systematic review and dose-response meta-analysis of prospective studies. *BMJ* **2016**, *353*, i2716. [CrossRef] [PubMed]
6. AOAC. Method 925.09, 991.43, 996.11, 986.11. In *Official Methods of Analysis*, 15th ed.; Association of Official Analytical Chemists: Arlington, VA, USA, 1996.
7. ISO/TS. *Food Products—Determination of the Total Nitrogen Content by Combustion According to the Dumas Principle and Calculation of the Crude Protein Content—Part 1 and 2: Cereals, Pulses and Milled Cereal Products (ISO/TS 16634-1 and ISO/TS 16634-2)*; International Organization for Standardization (ISO): Geneva, Switzerland, 2016.
8. AACC. *Approved Methods of AACC. Method 08-03, 30-10*, 9th ed.; The American Association of Cereal Chemists: St. Paul, MN, USA, 2000.
9. McKie, V.; McCleary, B.A. Novel and Rapid Colorimetric Method for Measuring Total Phosphorus and Phytic Acid in Foods and Animal Feeds. *J. AOAC Int.* **2016**, *99*, 738–743. [CrossRef]
10. Haros, M.; Rosell, C.M.; Benedito, C. Fungal phytase as a potential breadmaking additive. *Eur. Food Res. Technol.* **2001**, *213*, 317–322. [CrossRef]

Proceeding Paper

Effect of Quinoa Germination on Its Nutritional Properties †

Pedro Maldonado-Alvarado [1], Juan Abarca-Robles [1,2], Darío Javier Pavón-Vargas [1], Silvia Valencia-Chamorro [2] and Claudia Monika Haros [2,*]

1 Department of Food Science and Biotechnology, Escuela Politécnica Nacional, Quito 17-01-2759, Ecuador
2 Instituto de Agroquímica y Tecnología de Alimentos (IATA), Consejo Superior de Investigaciones Científicas (CSIC), Av. Agustín Escardino 7, Parque Científico, Paterna, 46980 Valencia, Spain
* Correspondence: cmharos@iata.csic.es

† Presented at the IV Conference Ia ValSe-Food CYTED and VII Symposium Chia-Link, La Plata and Jujuy, Argentina, 14–18 November 2022.

Abstract: The aim of this study was the evaluation of the effect of desaponification, soaking, germination and refrigerated storage on the phytase activity, phytic acid content, and nutritional properties of three varieties of quinoa: white, red and black. Desaponification and soaking reduced the amount of minerals and the nutritional content. Germination of the seeds was carried out in desaponified samples. Quinoa nutritional values, phytase activity and phytic acid content were measured during the first 7 days of germination, plus 7 days on refrigerated storage. Germination increased fibre and protein content as well as mineral contents. Germination significantly increased the phytase activity of all varieties and reduced the phytic acid content. The phytic acid content decreased during germination to between 32 and 74%. Refrigerated storage had no significant effect on most of the factors studied. Germination boosted nutritional content and phytase activity while decreasing phytic acid content. Germination can be a simple method to reduce phytic acid in quinoa and may also improve the nutritional quality of this pseudo-cereal, with potential for use in functional foods and vegetarian diets.

Keywords: *Chenopodium quinoa*; germination; phytic acid; phytase activity; nutritional value

Citation: Maldonado-Alvarado, P.; Abarca-Robles, J.; Pavón-Vargas, D.J.; Valencia-Chamorro, S.; Haros, C.M. Effect of Quinoa Germination on Its Nutritional Properties. *Biol. Life Sci. Forum* **2022**, *17*, 7. https://doi.org/10.3390/blsf2022017007

Academic Editors: Norma Sammán, Mabel Cristina Tomás and Loreto Muñoz

Published: 23 October 2022

Publisher's Note: MDPI stays neutral with regard to jurisdictional claims in published maps and institutional affiliations.

1. Introduction

Pseudocereal flours can be included in bakery products as a strategy to improve their nutritional profile without needing to use whole products completely [1]. The increasing interest in quinoa in Europe has generated a large number of studies with this seed as a partial substitute for refined wheat flour in bakery products as a strategy to improve their nutritional value. However, the wide genetic diversity of this seed offers very different compositions in different varieties [2]. However, quinoa, and other pseudocereals, has some anti-nutritional factors, especially saponins and phytic acid (phytates). Phytic acid can bind di- and trivalent minerals, making them unavailable for monogastric animals and humans. Phytate-degrading enzyme catalyses the hydrolysis of phytate, releasing the inorganic phosphate from the seeds. During the seed germination, the activity of several enzymes increase, including phytases [3].

The aim of this study was to evaluate the effect of germination on phytase activity on the proximate composition, mineral content and phytic acid residual of three quinoa varieties from Ecuador.

2. Materials and Methods

2.1. Materials

White, red and black quinoa harvested in Ecuador were the raw materials of this investigation. The samples were stored at room temperature and in a dark environment to prevent light exposure. Quinoa seeds were desaponified and disinfected prior to the germination process.

2.2. Germination Process

The soaking and sprouting processes were carried out according to the methodology described by D'ambrosio et al. [3].

2.3. Chemical Composition

The total lipid, fibre, ash and moisture contents of the samples were determined according to AOAC official methods: 945.16, 985.29, 923.03, and 925.10, respectively [4]. The protein content was analysed according to the Dumas method, whereas the carbohydrates were calculated based on the other measurements by difference. The determination of Ca, Fe and Zn was carried out by an atomic absorption spectrophotometer following the methodology described by Tazrart et al. [5]. Phytic acid content was measured according to the methodology described by Reason et al. [6], whereas the phytase-degrading enzyme activity as by Garcia-Mantrana et al. [7], expressed in U/g of quinoa. One phytase unit (U) was defined as 1.0 µg of inorganic phosphate liberated per minute at 50 °C and pH: 5.5.

The results were expressed as mean value ± standard deviation (SD). One-way ANOVA was performed to evaluate the statistical significance of differences.

3. Results

The protein content of the three studied quinoa seeds ranged between 16 and 19 %, which was in accordance with reported values in the bibliography [8]. The ash content of the quinoa samples ranged from 2.3 to 4.3 %, and lipids were between 7.0 and 7.9, in concordance with other researchers [9]. The mineral content values (Figure 1) obtained in this study were also similar to those described in the literature [10]. However, there was an initial reduction of ash content during soaking (data not shown) that could be explained by the mineral lixiviation during this step [9]. Later, during germination, the ash content significantly increased (data not shown), probably due to the conversion of carbohydrates to carbon dioxide during respiration as was observed previously by other researchers [11].

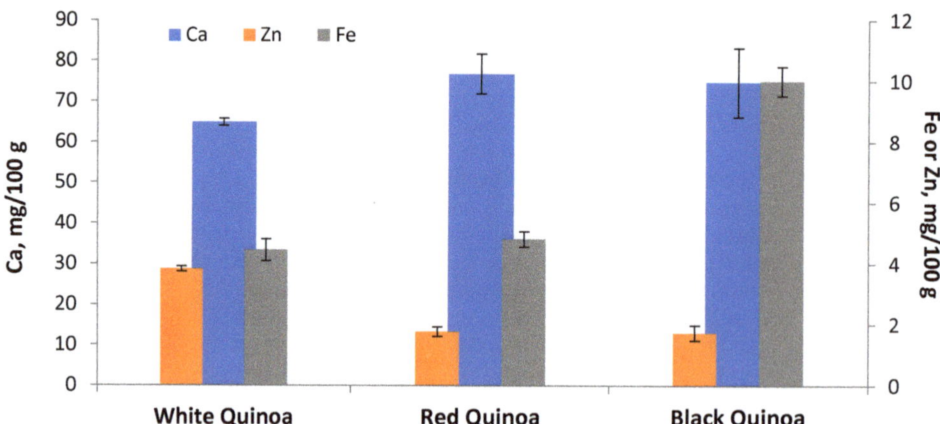

Figure 1. Mineral content in white, red and black quinoa from Ecuador.

The effects of germination on phytic acid content and phytase activity of the three quinoa varieties are presented in Figure 2. After one week of germinations, a significant reduction of phytic acid was observed in the three samples. In addition, phytase activity was increased in the first week of germination, with the exception of the black quinoa; however, the degradation of phytic acid was with the same efficiency as the other samples (Figure 2).

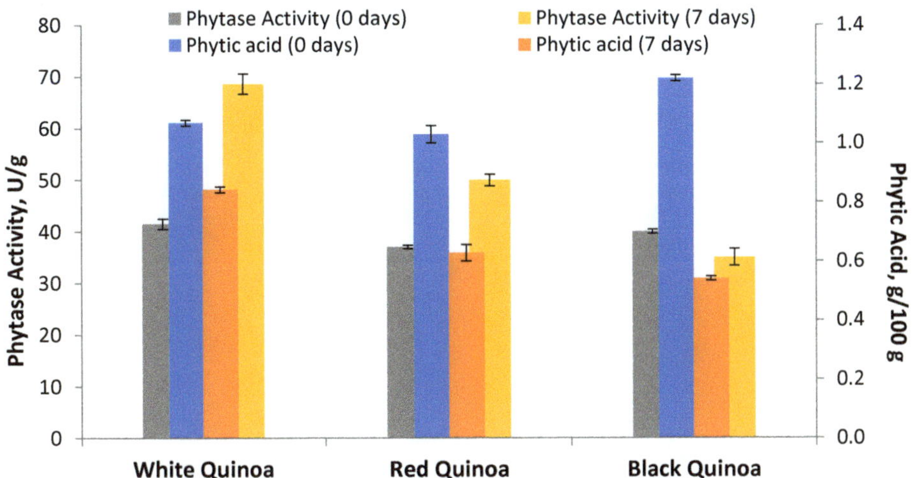

Figure 2. Phytic acid content and phytase activity in quinoa seeds before and after 7 days of germination.

4. Conclusions

The germination process has the potential to be an easy method to increase the mineral availability by phytate hydrolysis in the food production, mainly in vegetarian diets.

Author Contributions: Conceptualization, P.M.-A., S.V.-C. and C.M.H.; methodology, P.M.-A., S.V.-C. and C.M.H.; formal analysis, J.A.-R. and D.J.P.-V.; investigation, J.A.-R.; resources, P.M.-A., S.V.-C. and C.M.H.; writing—original draft preparation, D.J.P.-V. and J.A.-R.; writing—review and editing, D.J.P.-V., P.M.-A., S.V.-C. and C.M.H.; supervision, S.V.-C. and C.M.H.; project administration, P.M.-A. and C.M.H.; funding acquisition, P.M.-A. and C.M.H. All authors have read and agreed to the published version of the manuscript.

Funding: This work was supported by grant Ia ValSe-Food-CYTED (119RT0567), Food4ImNut (PID2019-107650RB-C21) from the Ministry of Sciences and Innovation (MICINN-Spain) and Escuela Politécnica Nacional, Quito, Ecuador Project PIJ 17-04 ("Elaboration of superfoods from Andean products").

Institutional Review Board Statement: Not applicable.

Informed Consent Statement: Not applicable.

Data Availability Statement: Not applicable.

Conflicts of Interest: The authors declare no conflict of interest.

References

1. Iglesias-Puig, E.; Monedero, V.; Haros, M. Bread with whole quinoa flour and bifidobacterial phytases increases dietary mineral intake bioavailability. *LWT Food Sci. Technol.* **2015**, *60*, 71–77. [CrossRef]
2. Lindeboom, N.; Chang, P.; Falk, K.; Tyler, R. Characteristics of starch from eight quinoa lines. *Cereal Chem.* **2005**, *82*, 216–222. [CrossRef]
3. D'Ambrosio, T.; Amodio, M.L.; Pastore, D.; De Santis, G.; Colelli, G. Chemical, physical and sensorial characterization of fresh quinoa sprouts (*Chenopodium quinoa* Willd.) and effects of modified atmosphere packaging on quality during cold storage. *Food Packag. Shelf Life* **2017**, *14*, 52–58. [CrossRef]
4. AOAC. *Official Methods of Analysis*, 18th ed.; Association of Official Analytical Chemists: Gaithersburgs, MD, USA, 2006.
5. Tazrart, K.; Lamacchia, C.; Zaidi, F.; Haros, M. Nutrient composition and in vitro digestibility of fresh pasta enriched with *Vicia faba*. *J. Food Compos. Anal.* **2016**, *47*, 8–15. [CrossRef]
6. Reason, D.; Watts, M.; Devez, A. Quantification of Phytic Acid in Grains. British Geological Survey. Open Report 18. 2015. Available online: https://nora.nerc.ac.uk/id/eprint/512947/ (accessed on 10 September 2022).
7. García-Mantrana, I.; Monedero, V.; Haros, M. *Myo*-inositol hexakisphosphate degradation by *Bifidobacterium pseudocatenulatum* ATCC 27919 improves mineral availability of high fibre rye-wheat sour bread. *Food Chem.* **2015**, *178*, 267–275. [CrossRef]

8. Martínez-Villaluenga, C.; Peñas, E.; Hernández-Ledesma, B. Pseudocereal grains: Nutritional value, health benefits and current applications for the development of gluten-free foods. *Food Chem. Toxicol.* **2020**, *137*, 111178. [CrossRef] [PubMed]
9. Pilco-Quesada, S.; Tian, Y.; Yang, B.; Repo-Carrasco-Valencia, R.; Suomela, J.P. Effects of germination and kilning on the phenolic compounds and nutritional properties of quinoa (*Chenopodium quinoa*) and kiwicha (*Amaranthus caudatus*). *J. Cereal Sci.* **2020**, *94*, 102996. [CrossRef]
10. Filho, A.M.M.; Pirozi, M.R.; Borges, J.T.D.S.; Pinheiro Sant'Ana, H.M.; Chaves, J.B.P.; Coimbra, J.S.D.R. Quinoa: Nutritional, functional, and antinutritional aspects. *Crit. Rev. Food Sci. Nutr.* **2017**, *57*, 1618–1630. [CrossRef] [PubMed]
11. Demir, B.; Bilgiçli, N. Changes in chemical and anti-nutritional properties of pasta enriched with raw and germinated quinoa (*Chenopodium quinoa* Willd.) flours. *J. Food Sci. Technol.* **2020**, *57*, 3884–3892. [CrossRef] [PubMed]

biology and life sciences forum

Proceeding Paper

Effect of Acid-Extrusion Cooking on Some Properties of Quinoa Starch †

Julio Rueda [1,2], Manuel Oscar Lobo [2], Norma Sammán [2,*] and Claudia Monika Haros [1,*]

1 Instituto de Agroquímica y Tecnología de Alimentos (IATA), Consejo Superior de Investigaciones Científicas (CSIC), Av. Agustín Escardino 7, Parque Científico, Paterna, 46980 Valencia, Spain
2 Centro Interdisciplinario de Tecnología y Desarrollo Social, CIITED-UNJu-CONCIET, Universidad Nacional de Jujuy, Ítalo Palanca 10, San Salvador de Jujuy 4600, Argentina
* Correspondence: normasamman@gmail.com (N.S.); cmharos@iata.csic.es (C.M.H.)
† Presented at the IV Conference Ia ValSe-Food CYTED and VII Symposium Chia-Link, La Plata and Jujuy, Argentina, 14–18 November 2022.

Abstract: According to the FAO, the economic potential of quinoa relies on the extraction and processing of its by-products. Starch in quinoa represents a major component. Although it has a limited application (due to its low solubility, high reactivity to hydrolysis or reactive hydroxyl groups), certain technological processes can modify or even improve the techno-functional or healthy properties. In this work, the effect of acid extrusion cooking on the molecular, chemical and morphological properties of quinoa starch was evaluated. A quinoa sub product from protein extraction (73% starch) was acid extruded (100 °C and 0, 10 and 40% of citric acid) and milled. Native (NS) and extruded (ES) samples were taken as the control. The resistant starch (RS) and free glucose (FG) content were measured though enzymatic methods. Molecular, structural and morphological characterization was assessed by infrared (IR) spectroscopy, particle size analysis by laser diffraction and optical microphotography. The results showed that an acid esterification at 40% caused a two-fold increase (1.10 g/100 starch dry basis) in the RS content, reduced the FG (mg/100 g db) from 801.36 (NS) to 368.56 and changed the IR spectrum due to the formation of new ester groups at a wavelength of 1712 cm^{-1} (carbonyl groups). Although, no significant differences were observed in the particle size distribution of the samples, microphotographs showed semi-crystalline structures (extruded and citrate starches) formed from native starch (starch aggregates). These data suggest that acid extrusion increased the RS content, formed citrate starch esters and changed the molecular and structural conformation of native quinoa starch. The evaluation of the additional properties would elucidate the effect of these changes on the bio and techno-functional properties.

Keywords: ester; extrusion; microestructure; quinoa; resistant starch; starch; glucose

Citation: Rueda, J.; Lobo, M.O.; Sammán, N.; Haros, C.M. Effect of Acid-Extrusion Cooking on Some Properties of Quinoa Starch. *Biol. Life Sci. Forum* **2022**, *17*, 8. https://doi.org/10.3390/blsf2022017008

Academic Editor: Mabel Cristina Tomás and Loreto Muñoz

Published: 24 October 2022

1. Introduction

According to the FAO, the potential of quinoa grains relies on procedures of extraction and the processing of quinoa by-products such as protein concentrates and starch, among others. Quinoa seeds have an important content of starch (55 to 60%, dry matter) which makes them appropriate as a carbohydrate source for processing and isolation [1]. Starch has food and non-food applications due to its versatile properties. It is a non-soluble biopolymer at room temperature, but instant starches have been prepared and developed to overcome these characteristics and make it even more useful for different applications [2]. From the nutrition point of view, starch not only provides the carbohydrates and energy necessary for human physiology, but it also contains starch forms that interact differently with the digestive system. Starch may be composed of rapidly digested starch, slowly digested starch and resistant starch (RS). Digestive enzymes do not hydrolyse RS and thus, they can be fermented in the colon and act as a source of dietary fibre. Among the four types of RS, RS4 is produced by chemical modifications and prevents the action of amylolytic enzymes by blocking their access [3]. In the colon, RS serves as a carbohydrate source for microbial fermentation and the production of short-chain fatty acids, which protect the colon from colorectal cancer. Additionally, a daily intake of food containing RS reduces the caloric value of meals and helps control the release of glucose into the bloodstream, thus reducing the insulin response [4]. Processing technologies for increasing the RS in foods such as cooking, baking, autoclaving and extrusion affect starch gelatinization and the retrogradation phenomena in starch, affecting the crystalline structure [5]. Extrusion is an interesting technique for the production of various food products since it has various advantages, such as being a continuously versatile process, having short processing times, a high yield and it is energetically efficient. Additionally, it can be combined with other treatments, such as different types of hydrolysis with enzymes and acids to give other products [6].

Changes produced during the extrusion process can affect the various properties of starchy materials. Important relations were described between the structural changes and rheological properties such as the texture, water binding capacity and viscoelastic characteristics of some food matrixes [2]. The objective of this study was to evaluate the effect of acid extrusion cooking on the molecular, chemical and structural changes in starch from quinoa seeds.

2. Materials and Methods

2.1. Quinoa Starch

Quinoa starchy material was obtained according to the procedure described by Rueda et al. [7]. The starchy pellet separated after the alkali protein extraction was air-dried (12 h, 40 °C), milled (0.5 mm of mesh) (Kinematica PX-MFC 90D, Bohemia, NY, USA) and stored in sealed-packed bags for a later use.

2.2. Acid Extrusion of Starch

The starch was prepared according to Ye et al. [4]. Aqueous solutions (100 mL) of citric acid at 0, 10 and 40% of the starch basis were mixed with 300 g of quinoa starch. Extrusion was performed in a KE19 Brabender (South Hackensack, NJ, USA) at a 3:1 compression ratio screw and 200 rpm, 26% moisture and 60, 70 and 100 °C for each barrel section.

2.3. Starch, Resistant Starch and Glucose Content Determination

The total and resistant starches were measured using the Megazyme Total Starch Assay Kit and the free glucose was measured using the Megazyme D-Glucose kit.

2.4. FTIR Spectrum of Starch Samples

The Fourier transforms infrared (FTIR) spectra of starch and modified starch samples were obtained using an FTIR spectrophotometer (FT/IR-4100, Jasco International Co. Ltd., Tokyo, Japan). The starch samples were in contact with the universal diamond ATR top-plate and each spectrum, from 4000–400 cm^{-1}, represented an average of the three scans. The absorbance readings were transformed into transmittance and plotted against the wavelength (cm^{-1}) using the software Origin (Origin Lab, Northampton, MA, USA).

2.5. Particle Size Distribution of Milled Starch

The particle size distribution in the samples was analysed by laser diffraction analysis (Malvern Instruments Ltd., Malvern, UK) equipped with an MS 15 Sample Presentation Unit. The particle size distribution was described by the following parameters: the largest particle size (D_{90}), the median diameter (D_{50}), the smallest particle size (D_{10}), Sauter mean diameter (D[3.2]) and mean particle diameter (D[4.3]). Triplicate measurements were performed at room temperature.

2.6. Optical Microphotography Characterization

The morphological characteristics of starch and the particle size of modified starch were evaluated by an Eclipse 90i Nikon wide-field microscope (Nikon Corporation, Tokio, Japan) equipped with a 5-megapixel cooled digital colour camera Nikon Digital Sight DS-5Mc (Nikon Corporation, Tokio, Japan). Reflected light images were achieved with an external USB-LED. The Nikon objective used was the CFI Plan Fluor 4X (MRH00040). The Samples were gently spread over a glass slide with a good particle separation. All microscopy images were acquired by using Nis-Elements Br 3.2 Software (Nikon Corporation, Tokio, Japan).

3. Results and Discussion

3.1. Changes in the Resistant Starch Content and Molecular Characterization

The effect of acid extrusion on the RS content is shown in Table 1. A slight decrease in the RS content was observed when the quinoa starch was extruded. However, the reactive extrusion with citric acid changed drastically the RS content and a reduction was observed at 10%, the extrusion at 40% of citric acid causing nearly a two-fold increase compared to the unextruded sample. A similar RS content was obtained by Neder-Suarez et al. [6] for acid extruded corn starch using a lower citric acid e than that employed in this study, and the moisture content seems to have an important role during the formation of RS. Hasjim and Jay-Lin Jane [8] found that acid-modified corn starch increased the RS amount by 2.5% after extrusion and attributed the increase to the formation of retrograded amylose.

Table 1. Resistant starch and free glucose content of native, extruded and acid extruded samples.

Parameter	QNS	QES0	QAES10	QAES40
Resistant starch (g/100 g starch)	0.63 ^{b} ± 0.08	0.56 ^{b} ± 0.07	0.20 ^{c} ± 0.04	1.10 ^{a} ± 0.09
Free glucose (mg/100 g sample)	801.36 ^{ab} ± 176.73	1064.37 ^{a} ± 50.15	904.62 ^{a} ± 88.31	368.56 ^{c} ± 6.76

Data represent mean ± standard deviation (n = 2). Values followed by different letters in each row mean statistical difference ($p < 0.05$). QNS: quinoa native starch. QES0: Quinoa extruded starch with no citric acid. QAES10: quinoa acid extruded starch with 10% of citric acid. QAES40: quinoa acid extruded starch with 40% of citric acid.

Free glucose (FG) content also was affected by the reactive extrusion process (Table 1). The extrusion caused a glucose liberation, suggesting a partial hydrolysis. The addition of citric acid diminished significantly the initial FG amount by more than two-fold times. The initial FG was reduced from an 801 to 368 mg/100 g sample after the addition of 40% of acid. These changes indicate not only the partial hydrolysis of starch during the extrusion, but also suggests the liberation of simple sugars and the readily available glucose for a reaction with citric acid.

The esterification reaction between starch and citric acid was assessed by an FTIR spectra of native starch, extruded starch, and acid extruded starch samples and changes in the characteristic functional groups were compared (Figure 1). All spectra had similar patterns and peak absorptions of quinoa starch and flour [1,4,9], however, when comparing the spectra of native and extruded starch against acid extruded starch, a new peak was observed at 1712 cm^{-1} in acid extruded samples and less intense peaks at 3300 and 2900, suggesting the esterification reaction took place in the native quinoa starch. Ye et al. [4] also observed a similar peak at 1730 cm^{-1} in citrate starch samples, after extrusion for carbonyl groups C=O, and they also observed less intense peaks around 3400 and 2930 in citrate starch samples, suggesting the formation of covalent groups between hydroxyl and carboxyl groups and starch molecules.

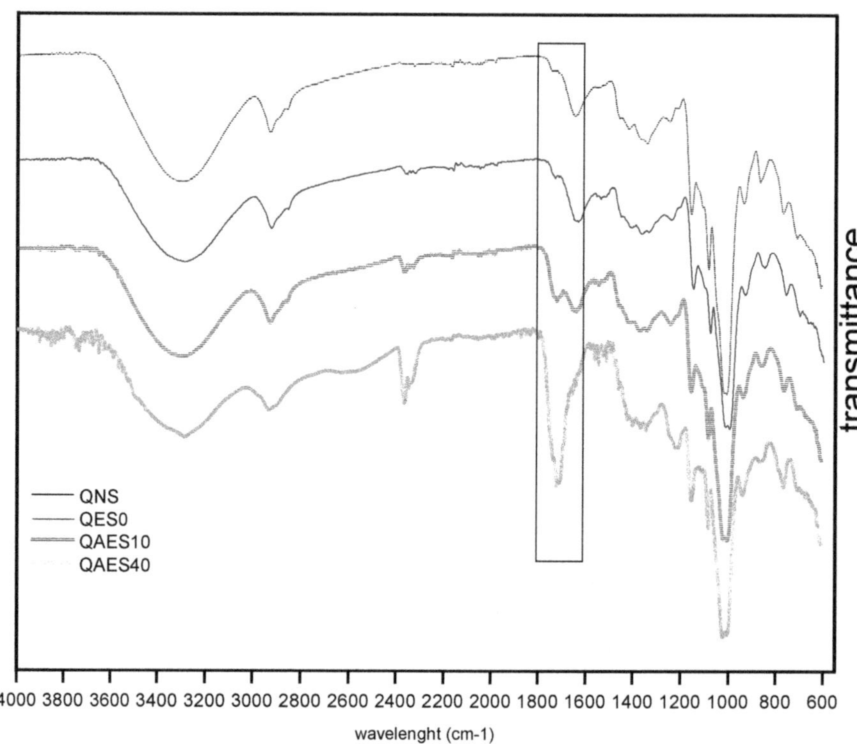

Figure 1. FTIR spectra of quinoa native starch (QNS); quinoa extruded starch (QES0); quinoa acid extruded starch 10% of citric acid (QAES10) and quinoa acid extruded starch at 40% of citric acid (QAES40).

3.2. Effect of Acid Extrusion on Particle Starch Properties

The behaviour and characteristics of food particulate materials, such as flours, depend on the geometry and size distribution of particles. In this study, light scattering and optical microscopy were used to evaluate the properties of native and processed quinoa starch. The particle size distributions of native and modified starch are shown in Table 2. The mean particle size D [4.3] of starch, extruded starch and acid extruded starch ranged from 282 to 321 µm. No significant differences were observed between native starch, extruded and acid extruded starch at 10%. Nevertheless, acid extrusion at 40% diminished significantly the mean diameter to 282.31 ± 4.70. The largest particle size ranged from 517 to 567 µm. Acid extrusion at 40% also diminished significantly ($p < 0.05$) the particle size.

Table 2. Particle size distribution of quinoa starch samples after extrusion and acid extrusion.

Parameter	Unit	Treatments			
		QNS	QES0	QAES10	QAES40
D_{10}	µm	26.81 ^{d} ± 19.12	90.67 ^{bc} ± 2.72	71.09 ^{c} ± 2.73	71.83 ^{c} ± 1.67
D_{50}	µm	319.45 ^{bc} ± 53.48	305.75 ^{bcd} ± 4.87	284.99 ^{bcd} ± 7.94	262.80 ^{d} ± 5.18
D_{90}	µm	560.72 ^{c} ± 4.04	567.63 ^{c} ± 5.47	558.56 ^{c} ± 6.10	517.84 ^{d} ± 6.05
D[3.2]	µm	45.99 ^{d} ± 18.79	145.53 ^{c} ± 10.91	132.78 ^{c} ± 5.17	136.071 ^{c} ± 2.54
D[4.3]	µm	295.8643 ^{c} ± 7.25	321.4907 ^{c} ± 4.09	303.55 ^{c} ± 6.39	282.3167 ^{cd} ± 4.70

Data represent mean ± standard deviation (n = 3). Values followed by different letters in each row mean statistical difference ($p < 0.05$). QNS: quinoa native starch; QES0: quinoa extruded starch with no citric acid; QAES10: quinoa acid extruded starch with 10% of citric acid; and QAES40: quinoa acid extruded starch with 40% of citric acid.

Although Table 2 shows similarities between the particle size distribution of quinoa starch flour, optical microphotography of the particulate material showed structural differences among the treatments. The microstructure of the particles is shown in Figure 2. All samples displayed particle diameters below 500 μm, however, starch aggregates can be observed in the native starch sample (Figure 2). Extruded (Figure 2 QES0) and acid extruded samples (Figure 2 QAES) showed a similar geometry but different than the particles observed for the native starch. Additionally, the translucent feature observed for extruded and acid extruded starch shows a clear difference and evidences an effect of citric acid on the starch particles. Figure 2 clearly shows that extrusion and reactive extrusion destroyed the starch aggregates and created amorphous structures. Neder-Suarez et al. [6] also found similar changes in the extruded samples of corn starch and attributed these changes to the gelatinization and dextrinization phenomena produced by a high pressure, temperature and mechanical forces during extrusion. Butt et al. [2] found similar results for rice extruded with citric acid using similar concentrations to the present study, and significant structural changes were reported.

Figure 2. Micrograph images of quinoa native starch (QNS); quinoa extruded starch (QES0); quinoa acid extruded starch 10% of citric acid (QAES10) and quinoa acid extruded starch at 40% of citric acid (QAES40).

4. Conclusions

The results showed that the reactive extrusion of quinoa starch increased the resistant starch content by producing cross-linking reactions between starch and citrate molecules, evidenced by the formations of ester groups, as revealed by the FTIR spectra. The two-fold increase in the resistant starch occurred at a concentration of 40% of citric acid. Acid extrusion cooking also changed the structural morphology of starch particles compared to native starch and significant three-dimensional conformational changes were shown by the micrographic analysis.

Author Contributions: Conceptualization, N.S. and C.M.H.; methodology, J.R. and M.O.L.; validation, N.S. and C.M.H.; formal analysis, J.R. and M.O.L.; investigation, J.R., M.O.L., N.S. and C.M.H.; resources, N.S. and C.M.H.; writing—original draft preparation J.R. and M.O.L.; writing—review and editing, N.S. and C.M.H.; visualization, J.R. and M.O.L.; supervision, N.S. and C.M.H.; project administration, N.S. and C.M.H.; funding acquisition, N.S. and C.M.H. All authors have read and agreed to the published version of the manuscript.

Funding: This work was supported by grant Ia ValSe-Food-CYTED (119RT0567), Secretaria de Ciencia, Tenica y Estudios Regionales Universidad Nacional de Jujuy and CONICET, and Food4ImNut (PID2019-107650RB-C21) from the Ministry of Sciences and Innovation (MICINN-Spain).

Institutional Review Board Statement: Not applicable.

Informed Consent Statement: Not applicable.

Data Availability Statement: Not applicable.

Conflicts of Interest: The authors declare no conflict of interest.

References

1. Contreras-jiménez, B.; Torres-vargas, O.L.; Rodríguez-garcía, M.E. Physicochemical characterization of quinoa (Chenopodium *quinoa*) flour and isolated starch. *Food Chem.* **2019**, *298*, 124982. [CrossRef] [PubMed]
2. Butt, N.A.; Ali, T.M.; Moin, A.; Hasnain, A. Comparative study on morphological, rheological and functional characteristics of extruded rice starch citrates and lactates. *Int. J. Biol. Macromol.* **2021**, *180*, 782–791. [CrossRef] [PubMed]
3. Dunrar, A.N.; Gocmen, D. Effects of autoclaving temperature and storing time on resistant starch formation and its functional and physicochemical properties. *Carbohydr. Polym.* **2013**, *97*, 764–771. [CrossRef]
4. Ye, J.; Luo, S.; Huang, A.; Chen, J.; Liu, C.; McClements, D.J. Synthesis and characterization of citric acid esterified rice starch by reactive extrusion: A new method of producing resistant starch. *Food Hydrocoll.* **2019**, *92*, 135–142. [CrossRef]
5. Masatcioglu, T.M.; Sumer, Z.; Koksel, H. An innovative approach for signi fi cantly increasing enzyme resistant starch type 3 content in high amylose starches by using extrusion cooking. *J. Cereal Sci.* **2017**, *74*, 95–102. [CrossRef]
6. Neder-Suárez, D.; Amaya-Guerra, C.A.; Pérez-Carrillo, E.; Quintero-Ramos, A.; Mendez-Zamora, G.; Sánchez-Madrigal, M.Á.; Lardizábal-Gutiérrez, D. Optimization of an extrusion cooking process to increase formation of resistant starch from corn starch with addition of citric acid. *Starch-Stärke* **2020**, *72*, 1900150. [CrossRef]
7. Rueda, J.; Lobo, M.O.; Sammán, N. Changes in the Antioxidant Activity of Peptides Released during the Hydrolysis of Quinoa (Chenopodium quinoa willd) Protein Concentrate. *Proceedings* **2020**, *53*, 12. [CrossRef]
8. Hasjim, J.; Jane, J.L. Production of resistant starch by extrusion cooking of acid-modified normal-maize starch. *J. Food Sci.* **2009**, *74*, C556–C562. [CrossRef] [PubMed]
9. Jan, K.N.; Panesar, P.S.; Singh, S. Process standardization for isolation of quinoa starch and its characterization in comparison with other starches. *J. Food Meas. Charact.* **2017**, *11*, 1919–1927. [CrossRef]

Proceeding Paper

Gluten-Free Couscous Made from Quinoa Sprouts: Study of Shelf Life †

Pedro Maldonado-Alvarado * and María Trujillo

Department of Food Science and Biotechnology, Escuela Politécnica Nacional, Quito P.O. Box 17-01-2759, Ecuador
* Correspondence: pedro.maldonado@epn.edu.ec

† Presented at the IV Conference Ia ValSe-Food CYTED and VII Symposium Chia-Link, La Plata and Jujuy, Argentina, 14–18 November 2022.

Abstract: The aim of this work was to determine the shelf life of a gluten-free couscous made from germinated quinoa. Desaponified quinoa of the Tunkahuan variety from Ecuador was used. Quinoa kernels were germinated, dried, milled, agglomerated, and finally steamed in controlled conditions. A designed particle agglomeration equipment was used to produce the couscous. The shelf life of the product was determined by accelerated testing. Product quality changes were evaluated during storage for 90 days in different types of packaging (cardboard, polyethylene polyester, and metallized polypropylene) under different conditions (15, 25, 35, and 45 °C). The moisture content, water activity, free fatty acids, and peroxide value showed a significant increase with time and temperature, while the maximum compression force showed a significant decrease. Analysis of total aerobes, total coliforms, molds, and yeasts showed that the product complies with the microbiological parameters established in the three types of packaging during storage. An increase in Aw and free fatty acids was found, whose kinetics of deterioration presented a first-order reaction. Aw activity was selected to estimate the shelf life of germinated quinoa couscous. Hence, the results suggest that this product can extend its shelf life at 20 °C up to 85 days and 136 days in cardboard and polyester polyethylene packaging, respectively.

Keywords: couscous; quinoa; germination; shelf life; accelerated testing

Citation: Maldonado-Alvarado, P.; Trujillo, M. Gluten-Free Couscous Made from Quinoa Sprouts: Study of Shelf Life. *Biol. Life Sci. Forum* **2022**, *17*, 9. https://doi.org/10.3390/blsf2022017009

Academic Editors: Norma Sammán, Mabel Cristina Tomás, Loreto Muñoz and Claudia Monika Haros

Published: 25 October 2022

1. Introduction

Couscous is a traditional staple food from North Africa, made of agglomerated granules from steamed wheat semolina. This product has been spread worldwide even using raw materials of botanical origin other than wheat for its manufacture, e.g., quinoa. Quinoa is a pseudocereal from the Andean region with high nutritional quality in protein, vitamins, and minerals, etc., and is considered a healthy food. Quinoa germination improves fiber and protein content, mineral availability, the concentration of total phenolic compounds, and phytase activity while decreasing phytic acid content [1]. To the knowledge of the authors, only one study has explored the effect of germination in quinoa to produce couscous. This work showed, that the nutritional and functional properties of germinated quinoa couscous were maximized compared to couscous from ungerminated quinoa and traditional couscous [2]. However, its shelf life has not been studied.

The objective of this work was to determine the shelf life by accelerated shelf-life testing of a couscous made from Ecuadorian fermented quinoa stored in different packaging (cardboard, polyethylene polyester, and metallized polypropylene).

2. Materials and Methods

2.1. Raw Material

Desaponified quinoa of the Tunkahuan variety (golden quinoa) from Ecuador was used. Quinoa kernels were disinfected and stored in a chamber at room temperature, in the dark.

2.2. Germination

After soaking, quinoa was placed into a germinator for 24 h at 20 °C and 100% RH.

2.3. Couscous Preparation

Quinoa sprouted, was dried in an oven, then ground and sifted, agglomerated, and finally steamed and dried in controlled conditions. A designed particle agglomeration equipment was used to produce the couscous [2].

2.4. Shelf-Life

The shelf life of the product was determined by accelerated testing. A mixed experimental design was performed to determine the shelf life of germinated quinoa couscous stored in different packaging. Product quality changes were evaluated during storage for 3 months in different types of packaging (cardboard, polyethylene polyester, and metallized polypropylene) under different conditions (15, 25, 35, and 45 °C). The moisture content, water activity, free fatty acids, peroxide value, and maximum compressive strength were evaluated as well as microbiological analysis i.e., total aerobes, total coliforms, molds, and yeasts [3].

The results of each of the factors were evaluated by means of an analysis of variance (ANOVA) and a least significant difference test by Fisher's LSD method, with a confidence level of 95% ($p < 0.05$).

3. Results

The determination of the kinetics of deterioration was obtained in a short period of time using accelerated tests. For which, the germinated quinoa couscous was stored in cardboard packaging and at high temperatures, conditions that accelerated its deterioration process, which allowed the increase of unwanted factors. During the 90 days of storage of the germinated quinoa couscous at all the temperatures evaluated, it did not reach the maximum permissible limit (MAL) of moisture content (13.5%) [4]. In water activity, the same previous tendency was observed, the higher temperature, the decrease in water activity, and the lower time. However, it exceeded the MAL of water activity 0.6% [5] in 15 and 25 °C from 75 days of storage. This could be explained by the permeability of cardboard packaging. Storage of the product in a place with a relative humidity greater than that of equilibrium allowed the migration of water vapor toward the food. For germinated quinoa couscous stored in cardboard, the deterioration kinetics due to water activity was adjusted to a reaction order one. The increase in free fatty acids is expressed as a percentage of oleic acid as a function of time at the different studied temperatures for germinated quinoa couscous stored in cardboard. It exceeded the MAL of free fatty acids by 0.5% in all studied temperatures at 90 days of storage. The increase in fatty acids can be caused by the enzymatic hydrolysis of lipids, which is also related to the increase in humidity [6].

For germinated quinoa couscous stored in cardboard, the deterioration kinetics due to free fatty acids were adjusted to a first-order reaction. The peroxide index at the end of the storage period at 45 °C increased. The maximum value obtained is less than the maximum permitted peroxide index limit (10 meq O_2/kg oil) [7]. On the other hand, storage period and storage temperature had a significant effect on the peroxide value. The presence of the peroxide index indicates the production of peroxides and the increase of short-chain fatty acids that can cause oxidative rancidity in the food. The storage time and temperature of couscous made from germinated quinoa have significant effects on the maximum compression force (10 kg force for foods). Changes in compressive force can be caused by moisture absorption on the surface of the food, enzymatic hydrolysis, and other physical and chemical deterioration [8]. Water activity [9] and free fatty acids were the determinant factors of shelf-life analyses in this couscous type, due to the results obtained exceeded the maximum permissible limits. Regarding water activity, from the constants obtained from the deterioration kinetics, an activation energy of 4.83 kcal/mol, and a determination coefficient of 0.989 were obtained. With the data obtained and for a maximum permissible water activity of 0.6, a useful life of 85 days was obtained. For free fatty acids, an activation energy of 1.65 kcal/mol, and a determination coefficient of 0.978 were determined for the cardboard packaging. With the results obtained and with a permissible limit of 0.5% oleic acid, a shelf life of 97 days was obtained.

Regarding microbiological analyses, total aerobes, total coliforms, molds, and yeasts were not detected during the 90-day storage period. This, despite the increased moisture content and water activity, indicates that sprouted quinoa couscous is microbiologically safe.

The results of the multiple response optimization to determine the effect of packaging material on the shelf life of germinated quinoa couscous showed that water activity was the most important quality index for determining the kinetic model of deterioration.

In this way, the shelf-life values of gluten-free couscous made from germinated quinoa were defined. The estimated shelf life for the cardboard, polyethylene polyester, metallized polypropylene packaging was 85, 136, and 86 days and for the packaging, 136 days, which means that after this storage time, the quinoa couscous would no longer meet the quality limits.

4. Conclusions

During the storage of germinated quinoa couscous, the change in water activity and fatty acids presented a reaction order of 1. Water activity was the determinant factor selected to estimate the shelf life of the germinated quinoa couscous. Germinated quinoa couscous at the end of the 90-day was considered microbiologically safe. The best packaging for storing gluten-free couscous from germinated quinoa was polyethylene polyester packaging.

Author Contributions: Conceptualization, P.M.-A.; Formal analysis, P.M.-A. and M.T.; Methodology, P.M.-A. and M.T.; Supervision, P.M.-A.; Validation, P.M.-A.; Writing—original draft, P.M.-A.; Writing—review and editing, P.M.-A. All authors have read and agreed to the published version of the manuscript.

Funding: This work was supported by PIJ 17-04 research project "Elaboration of new nutritional products from Andean superfoods" funded by the Escuela Politécnica Nacional and the Department of Food Science and Biotechnology. The APC was funded by IA ValSe-Food-CYTED (119RT0567), Food4ImNut (PID2019-107650RB-C21) from the Ministry of Sciences and Innovation (MICINN-Spain).

Institutional Review Board Statement: Not applicable.

Informed Consent Statement: Not applicable.

Data Availability Statement: The data analyzed in this study are available from the authors upon reasonable request.

Acknowledgments: The authors acknowledge the financial support of PIJ 17-04 research project "Elaboration of new nutritional products from Andean superfoods" funded by the Escuela Politénica Nacional and the Department of Food Science and Biotechnology and IA ValSe-Food-CYTED (119RT0567).

Conflicts of Interest: The authors declare no conflict of interest.

References

1. Pilco-Quesada, S.; Tian, Y.; Yang, B.; Repo-Carrasco-Valencia, R.; Suomela, J.P. Effects of germination and kilning on the phenolic compounds and nutritional properties of quinoa (Chenopodium quinoa) and kiwicha (Amaranthus caudatus). *J. Cereal Sci.* **2020**, *94*, 102996. [CrossRef]
2. Maldonado-Alvarado, P.; Nicolalde, K. Evaluation of the quality properties of couscous made from germinated quinoa. (Abstr.). Cereals & Grains 20 Online. *Cereal Chem.* **2021**, *98*, S1–S37. [CrossRef]
3. Torrieri, E. Storage Stability: Shelf Life Testing. In *Encyclopedia of Food and Health*; Caballero, B., Finglas, P.M., Toldrá, F., Eds.; Academic Press: Oxford, UK, 2016; pp. 188–192. [CrossRef]
4. D'ambrosio, T.; Amodio, M.L.; Pastore, D.; de Santis, G.; Colelli, G. Chemical, physical and sensorial characterization of fresh quinoa sprouts (Chenopodium quinoa Willd.) and effects of modified atmosphere packaging on quality during cold storage. *Food Packag. Shelf Life* **2017**, *14*, 52–58. [CrossRef]
5. Rahman, M.S. Water Activity and Glass Transition of Foods. *Ref. Modul. Food Sci.* **2019**, 1–10. [CrossRef]

6. Nantanga, K.K.M.; Seetharaman, K.; de Kock, H.L.; Taylor, J.R.N. Thermal treatments to partially pre-cook and improve the shelf-life of whole pearl millet flour. *J. Sci. Food Agric.* **2008**, *88*, 1892. [CrossRef]
7. Kong, F.; Singh, R.P. Stab. Chemical Deterioration and Physical Instability of Foods and Beverages. *Shelf Life Food* **2016**, *43*, 43–76. [CrossRef]
8. de Bouillé, A.G.; Beeren, C.J.M. Sensory Evaluation Methods for Food and Beverage Shelf Life Assessment. In *The Stability and Shelf Life of Food*, 2nd ed.; Subramaniam P, Ed.; Woodhead Publishing: Philadelphia, PA, USA, 2016; pp. 199–228. [CrossRef]
9. Pfeiffer, C.; D'aujourd'hui, M.; Walter, J.; Nuessli, J.; Escher, F. Optimizing food packaging of bakery products. *Food Technol.* **1999**, *53*, 52–59.

Proceeding Paper

Chia Oil Microencapsulation by Spray Drying Using Modified Soy Protein as Wall Material †

Paola Alejandra Gimenez [1], Antonella Estefanía Bergesse [2], Nahuel Camacho [3], Pablo Daniel Ribotta [4], Marcela Lilian Martínez [5] and Agustín González [1,*]

1 Faculty of Chemistry, National University of Córdoba, IPQA-CONICET, Córdoba 5000, Argentina
2 Faculty of Agricultural Sciences, National University of Córdoba, IMBIV-CONICET, Córdoba 5000, Argentina
3 Faculty of Chemistry, National University of Córdoba, UNITEFA-CONICET, Córdoba 5000, Argentina
4 Faculty of Exact, Physical and Natural Sciences, National University of Córdoba, ICITAC-CONICET, Córdoba 5000, Argentina
5 Faculty of Exact, Physical and Natural Sciences, National University of Córdoba, IMBIV-CONICET, Córdoba 5000, Argentina
* Correspondence: agustingonzalez@unc.edu.ar
† Presented at the IV Conference Ia ValSe-Food CYTED and VII Symposium Chia-Link, La Plata and Jujuy, Argentina, 14–18 November 2022.

Abstract: Chia seed is the richest vegetable source of polyunsaturated fatty acids. A diet rich in these fatty acids decreases the risk of many chronic non-communicable diseases. The incorporation of omega-3-rich oils in processed foods seems to be an efficient way to increase its consumption; however, when these are exposed to processing conditions, oxidation reactions occur. Microencapsulation technology is an alternative method to enhance lipid stability. The use of vegetable proteins as wall materials is being widely developed. The objective of the present work was to study the effect of the microencapsulation using chemically modified soy protein as wall material on chia oil oxidative stability. The crosslinking effect on soy protein with different concentrations of gallic and tannic acids was evaluated. Chia oil was incorporated into the dispersions using a high-speed homogenizer and the emulsions were dried by Spray-Drying. The microcapsule moisture content and water activity were around 2.96–5.86% and 0.17, respectively. The encapsulation efficiency was among 54–78%. The oxidative stability determined by the Rancimat analysis showed a positive correlation between the amount of cross-linking agent used and the induction time, reaching a maximum of 11.97 h. In a storage test, the peroxide value was markedly lower for those crosslinked microcapsules respect to pure SPI wall material after 90 days. The results demonstrated that use of these polyphenols as crosslinkers of the wall material exerts a positive effect in the protection of the chia oil derived from obtaining an optimized wall material and from the intrinsic antioxidant properties of these crosslinkers.

Keywords: antioxidants; chia oil; crosslinking; microencapsulation; polyphenols

Citation: Gimenez, P.A.; Bergesse, A.E.; Camacho, N.; Ribotta, P.D.; Martínez, M.L.; González, A. Chia Oil Microencapsulation by Spray Drying Using Modified Soy Protein as Wall Material. *Biol. Life Sci. Forum* **2022**, *17*, 10. https://doi.org/10.3390/blsf2022017010

Academic Editors: Norma Sammán, Mabel Cristina Tomás, Loreto Muñoz and Claudia Monika Haros

Published: 25 October 2022

1. Introduction

Chia seed oil with 61–70% of alpha linolenic acid (18:3) is the richest vegetable source of omega-3 fatty acids. The consumption of fatty acids of the omega 3 series provides numerous health benefits and can be incorporated in the form of triglycerides or ethyl esters [1]. A diet rich in these compounds reduces the risk of contracting coronary and neurodegenerative diseases, cancer, metabolic syndrome, rheumatism, type 2 diabetes, atherosclerosis, and Alzheimer's disease [1]. Although a higher consumption of these lipids is favorable from a nutritional and healthy point of view, some drawbacks are present due to its less oxidative stability and a short shelf life. One of the main challenges for the use and incorporation of these fatty acids in processed foods lies in the need to be stabilized by incorporating antioxidants and conveyed in a polymeric matrix that contains and protects them. An effective technology for this purpose is microencapsulation in solid and polymeric matrices [2]. The use of plant proteins as wall material is being widely developed; however, it is necessary to study strategies that make it possible to obtain better microcapsules compared to those reported until now [3].

The antioxidant properties of polyphenols such as tannic and gallic acid have been reported by Shavandi et al. [4]. In addition, Kim et al. [5] reported an increase in the antioxidant properties of tannic acid after a thermal treatment.

Therefore, the objective of the present work was to study a challenging and novel strategy such as the chemical modification of the proteins used as wall material through cross-linking reactions with natural polyphenolic compounds. The influence on the chemical quality of the omega 3 rich oils containing microcapsules (MC) was analyzed, focusing on their degree of protection of the contained oil.

2. Materials and Methods

2.1. Materials

Soy protein isolate (SPI) (SUPRO E, 90% protein, fat-free dry basis) was obtained from DuPont Nutrition & Health (EEUU) and chia oil (CO) was obtained from seeds from the province of Salta, Argentina (Distribuidora Nicco SRL, Córdoba, Argentina). The reagents and solvents used were purchased from local distributors.

2.2. Oil Extraction and Microcapsule Preparation

Chia oil (CO) was extracted by cold pressing in a single step with a Komet screw-press (CA 59 G model, IBG Monforts, Mönchengladbach, Germany) according to Martinez et al. [6].

Aqueous dispersions of isolated soy protein (SPI) 8% *w*/*w* were prepared and brought to pH between 9 and 11. Natural crosslinkers such as tannic (TA), gallic acid (GA) and a heat-treated tannic acid (130 °C in autoclave for 15 and 30 min for increasing its antioxidant capacity [5]) (TA15 and TA30) were studied. Different amounts of the crosslinking agents were added (1–10% *w*/*w* respect to SPI amount) and allowed to react with stirring at 60 °C for 24 h. CO was incorporated dropwise into the dispersions at a 2:1 ratio (SPI:oil) for 15 min at 18,000 rpm using an Ultraturrax homogenizer (IKA T18, Staufen, Germany). The resulting emulsions were dried in a Mini Spray Dryer Büchi B-290 (Büchi Labortechnik AG, Flawil, Switzerland) with a two-fluid nozzle under the following conditions: air inlet and outlet temperature 130 and 80 °C, respectively, pump 10%, aspirator 100% and air flow 538 L/h.

2.3. Microcapsule Characterization

Moisture content was measured for each sample with a moisture analyzer with halogen heating (model HE53 Mettler Toledo, Columbus, OH, United States). Water activity was measured with Aqua-Lab (208 Series 3, Decagon Devices Inc., Pullman, WA, USA) at 25.0 ± 0.5 °C. Surface or free oil (SO), total oil (TO) and encapsulation efficiency (EE) were assessed following a previously reported methodology [2].

2.4. Oil Oxidative Stability Study

To study the oxidative stability of unencapsulated and encapsulated oil, samples were subjected to accelerated oxidation conditions (100 °C, air flow 20 L/h) in a Rancimat apparatus (METROHM, Herisau, Switzerland) and expressed as induction period (IP). The protection factor (PF) was defined as the ratio of IP of the microencapsulated oil and IP of unencapsulated oil. The hydroperoxide values (HPV) were assessed following the methodologies of González et al. [2].

The fatty acid composition of oil was analyzed by gas chromatography according to what was proposed by González et al. [7]. For the storage stability test, bulk CO and microencapsulated oil were placed in 250 mL amber glass bottles in a thermostated chamber at 25 °C. The samples were stored for 90 days. At different times, 5 g of samples were extracted, and chia oil was extracted by immersing them in hexane for 24 h at 4 °C and evaporating the solvent in vacuum at 36 °C in order to evaluate their hydroperoxide value (HPV).

2.5. Statistical Analysis

Analytical determinations were the averages of duplicate measurements. Statistical differences among treatments were estimated from ANOVA test at the 5% level ($p < 0.05$) of significance, for all parameters evaluated.

3. Results and Discussion

3.1. Microcapsule Characterization

Dark green powders were obtained by spray-drying the prepared emulsions. The color of the powders darkened with increasing the amount of crosslinker initially added. The microcapsule moisture content and water activity were around 2.96 and 5.86% and 0.17, respectively, which is similar to reported in the literature (Bordon et al. [8]). With regard to the oil distribution in the microcapsules, the encapsulation efficiency was between 54 and 78%. Similar results were also obtained by González et al. [2] for SPI and maltodextrin microcapsules with values between 52 and 65%; Bordon et al. [8] for SPI and gum Arabic microcapsules with values between 68 and 87%. The relative abundance of the unsaturated fatty acids after oil microencapsulation was composed of 64.1%, 19.4% and 6.5% of linolenic, linoleic, and oleic acid, respectively. These results were in concordance with those reported in the literature (Martinez et al. [9]) for press-cold chia oil.

The protective effect of the microcapsules wall materials on the oxidative stability of chia oil was demonstrated by obtaining high induction periods in the Rancimat test. Microcapsules with wall materials with 1%, 5% and 10% of crosslinker agent showed IPs of 7.6 h, 7.6–11.9 h and 6.7–11.0 h, respectively. The maximum protection factor (PF) was around 4.0 for 5% of crosslinker agents. These results were higher than those reported by González et al. [2] with a maximum of 6.4 h and a PF of 2.7. In addition, the antioxidant properties of the natural crosslinkers were assessed adding the same amount of the polyphenols in bulk oils. The results of IP and PF for bulk chia oil with 5% of polyphenols were 5.6 h and 2.3, respectively. These results show that polyphenols exert an antioxidant effect, but this effect is smaller in respect to the protection obtained for microencapsulated oil. It could be seen that the simple addition of both effects in separately as microencapsulation (sample with microencapsulated oil in SPI wall material without crosslinker) and antioxidant effect (bulk oil with polyphenols samples) is smaller than both strategies combined (microencapsulated oil in crosslinked SPI wall material). This observation allows us to determine that there is a synergistic relationship between both protection strategies.

3.2. Storage Test

For this assay, microcapsules with 5% of crosslinker agent were selected due to their higher PF with the lower percentage of polyphenols used. Figure 1 shows that hydroperoxide values (HPV) gradually increased with storage time. After 90 days, all the samples analyzed presented HPV higher than the Codex acceptance limit for cold-pressed and virgin vegetable oils (15 milliequivalents of O_2/kg oil) (Codex Alimentarius, 2011). It is important to highlight that the microencapsulated samples whose wall material was composed by cross-linked protein presented a significantly lower HPV ($p < 0.05$) than the microcapsules with protein without chemical modification as wall material. The difference in oxidative stability of the microencapsulated chia oil with SPI in the present study compared with González et al. [2] could be due to the absence of the heating stage in the wall material preparation in which Maillard reaction occurs and the resulting compounds have shown antioxidant activity [10].

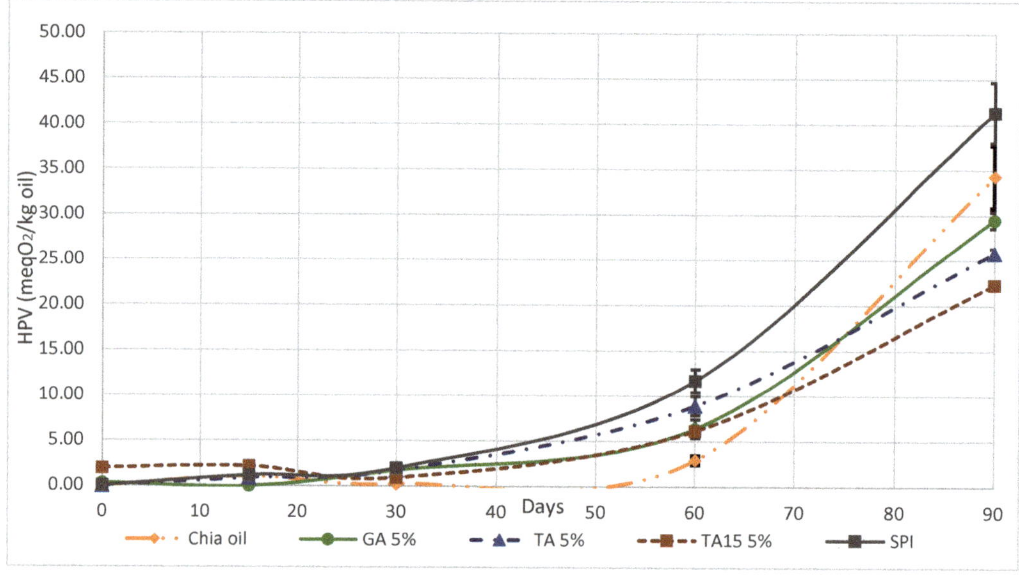

Figure 1. The peroxide value evolution for different storage times.

4. Conclusions

The results demonstrated that the use of these polyphenols as crosslinkers of the wall material for microencapsulation of chia oil exerts a positive effect on the protection of the oil derived from obtaining an optimized wall material and from the intrinsic antioxidant properties of these crosslinkers, allowing its potential application as an ingredient in processed foods.

Author Contributions: Conceptualization, M.L.M. and A.G.; methodology, P.A.G., M.L.M. and A.G.; validation, M.L.M. and A.G.; investigation, P.A.G., A.E.B. and N.C.; writing-original draft preparation, P.A.G.; writing-review and editing, M.L.M. and A.G.; supervision M.L.M. and A.G.; project administration P.D.R.; funding acquisition P.D.R. All authors have read and agreed to the published version of the manuscript.

Funding: This research was founded by La ValSe-Food-CYTED, grant number 119RT0567, FONCyT grant number PICT–2019-04265-II-B.

Institutional Review Board Statement: Not applicable.

Informed Consent Statement: Not applicable.

Data Availability Statement: Not applicable.

Conflicts of Interest: The authors declare no conflict of interest.

References

1. Racine, R.A.; Deckelbaumb, R.J. Sources of the very-long-chain unsaturated omega-3 fatty acids: Eicosapentaenoic acid and docosahexaenoic acid. *Curr. Opin. Clin. Nutr. Metab. Care* **2007**, *10*, 123–128. [CrossRef] [PubMed]
2. González, A.; Martínez, M.L.; Paredes, A.J.; León, A.E.; Ribotta, P.D. Study of the preparation process and variation of wall components in chia (*Salvia hispanica* L.) oil microencapsulation. *Powder Technol.* **2016**, *301*, 868–875. [CrossRef]
3. Heck, R.T.; Lorenzo, J.M.; Dos Santos, B.A.; Cichoski, A.J.; de Menezes, C.R.; Campagnol, P.C.B. Microencapsulation of healthier oils: An efficient strategy to improve the lipid profile of meat products. *Curr. Opin. Food Sci.* **2020**, *40*, 6–12. [CrossRef]
4. Shavandi, A.; Bekhit, A.E.D.A.; Saeedi, P.; Izadifar, Z.; Bekhit, A.A.; Khademhosseini, A. Polyphenol uses in biomaterials engineering. *Biomaterials* **2018**, *167*, 91–106. [CrossRef] [PubMed]
5. Kim, T.J.; Silva, J.L.; Kim, M.K.; Jung, Y.S. Enhanced antioxidant capacity and antimicrobial activity of tannic acid by thermal processing. *Food Chem.* **2010**, *118*, 740–746. [CrossRef]

6. Martínez, M.L.; Marín, M.A.; Salgado Faller, C.M.; Revol, J.; Penci, M.C.; Ribotta, P.D. Chia (*Salvia hispanica* L.) oil extraction: Study of processing parameters. *LWT—Food Sci. Technol.* **2012**, *47*, 78–82. [CrossRef]
7. González, A.; Martínez, M.L.; León, A.E.; Ribotta, P.D. Effects on bread and oil quality after functionalization with microencapsulated chia oil. *J. Sci. Food Agric.* **2018**, *98*, 4903–4910. [CrossRef] [PubMed]
8. Bordón, M.G.; Bodoira, R.M.; González, A.; Piloni, R.; Ribotta, P.D.; Martínez, M.L. Spray-Drying, Oil Blending, and the Addition of Antioxidants Enhance the Storage Stability at Room Temperature of Omega-3-Rich Microcapsules Based on Chia Oil. *Eur. J. Lipid Sci. Technol.* **2022**, *124*, 1–11. [CrossRef]
9. Martínez, M.L.; Curti, M.I.; Roccia, P.; Llabot, J.M.; Penci, M.C.; Bodoira, R.M.; Ribotta, P.D. Oxidative stability of walnut (*Juglans regia* L.) and chia (*Salvia hispanica* L.) oils microencapsulated by spray drying. *Powder Technol.* **2015**, *270*, 271–277. [CrossRef]
10. Copado, C.N.; Diehl, B.W.K.; Ixtaina, V.Y.; Tomás, M.C. Application of Maillard reaction products on chia seed oil microcapsules with different core/wall ratios. *LWT—Food Sci. Technol.* **2017**, *86*, 408–417. [CrossRef]

biology and life sciences forum

Proceeding Paper

Obtaining a Functional Food from Andean Grains through Lactic Acid Fermentation †

Cintya Elizabeth Salinas Alcon, María Dolores Jiménez, Manuel Oscar Lobo and Norma Cristina Sammán *

CIITED—Centro de Investigación Interdisciplinario en Tecnología y Desarrollo Social del NOA—CONICET, Facultad de Ingeniería, Universidad Nacional de Jujuy, Italo Palanca 10, San Salvador de Jujuy 4600, Argentina
* Correspondence: normasamman@gmail.com
† Presented at the IV Conference Ia ValSe-Food CYTED and VII Symposium Chia-Link, La Plata and Jujuy, Argentina, 14–18 November 2022.

Abstract: Quinoa and amaranth have excellent nutritional compositions. Lactic fermentation is capable of transforming the functional, structural, organoleptic and nutritional properties of raw materials. The objective of this study was to develop a product analogous to yogurt, which would be suitable for special diets. The product was formulated with quinoa and amaranth flours, water, sugar and strawberries, and it was fermented with an exopolysaccharide-producing strain of Lactobacillus Plantarum. Chemical parameters and BAL growth were monitored. Sensory analysis determined the best formulation and fermentation time. In the final product, the proximal composition, microbial count, pH, antioxidant activity, color, viscosity and content of exopolysaccharides (EPS) were determined. The formulation of the selected product was 15 g quinoa/amaranth (50:50) flour, 12 g sugar, 25 g strawberry pulp and 85 g water, fermented for 8 h. The composition of the functional product was 19.60 g carbohydrates and 1.74 g protein/100 g of the puree. The viable cell count was 7.60×10^8 CFU/g; the pH was 3.86; and IC50 = 10.3 mg/mL. The color parameters L*, a* and b* were 43.85, 15.24 and 11.72, respectively, with a reddish-brown color. During fermentation, the viscosity increased to 5029 mPa*s at 10 rpm due to the production of EPS (6.78 g EPS/L fermented pure). However, EPS production was not enough to significantly modify the viscosity, probably due to the amylolytic capacity of BAL. Fermented pure was described as having a rich, fruity and acidic flavor, with a mild and pleasant smell and a viscous texture. The food obtained was analogous to yogurt with acceptable sensory characteristics and was suitable for vegetarians, coeliacs and lactose intolerants.

Keywords: amaranth; fermentation; functional food; Lactobacillus Plantarum; quinoa

Citation: Salinas Alcon, C.E.; Jiménez, M.D.; Lobo, M.O.; Sammán, N.C. Obtaining a Functional Food from Andean Grains through Lactic Acid Fermentation. *Biol. Life Sci. Forum* **2022**, *17*, 11. https://doi.org/10.3390/blsf2022017011

Academic Editors: Mabel Cristina Tomás, Loreto Muñoz and Claudia Monika Haros

Published: 25 October 2022

1. Introduction

In recent years, the demand for functional foods and beverages has increased globally due to consumer trends seeking functional, organic, nutraceutical and allergen-free foods, among other characteristics. Consequently, the food and beverage industry has researched new sources of raw materials, new technologies and new food processes in order to develop innovative products that meet these needs, for example, plant-based dairy analogues [1]. The health problems associated with dairy consumption, the growing trend of veganism and the limited availability of these products in many countries have led to the development of dairy-free probiotic foods. Andean grains are excellent substrates to produce fermented functional foods since they are complete matrices with adequate profiles of macro and micronutrients, which allow the growth of nutritionally demanding lactic acid bacteria [2]. Quinoa (*Chenopodium quinoa*) and amaranth (*Amaranthus* spp.) have a high nutritional value without lactose, gluten and cholesterol. The aim of this work was to develop a product analogous to yogurt through the lactic fermentation of quinoa and amaranth.

2. Materials and Methods

2.1. Raw Materials

Quinoa (variety INTA Hornillos) and amaranth (variety Cica) were obtained from the *Centro de Investigación y Desarrollo Tecnológico para la Agricultura Familiar* IPAF NOA (Research and Technological Development Center for Family Agriculture INTA Hornillos, Jujuy, Argentina). The saponin of the quinoa was removed with successive washes. Both grains were washed and dried in a forced convection oven at 40 °C and then milled to obtain flours.

Lactic bacteria (Lactobacillus Plantarum), producer of exopolysaccharides, was used. The starter culture was prepared with activation in MRS broth and was incubated at 36 °C for 24 h. Then it was replicated twice in MRS broth at concentration of 1% and incubated for 12 and 6 h, respectively, to reach the exponential phase.

2.2. Sample Preparation

Different amounts of quinoa and amaranth flours (50:50 ratio) were studied to select the optimal ratio of flour–water (between 5 and 30% *w/v* of flour) to obtain the product. The addition of sugar (12 and 15 g/100 g puree) and strawberry pulp (25 and 30 g/112 g puree) were also studied to choose the formulations with the best physicochemical and sensory characteristics.

According to the previous results, the puree was made by mixing 15 g of quinoa/amaranth flour (50:50), 12 g of sugar and 85 g of water. Sugar was used to promote exopolysaccharide (EPS) synthesis [3] and to improve flavor. The puree was cooked in a water bath for 10 min to cook and gelatinize the starch to obtain a suitable viscous texture and to avoid phase separation. Separately, the strawberry was processed to obtain a pulp and added to the cooked puree. The puree was packed in glass flasks with screwed-on metal caps and autoclaved (P = 0.9 kg/cm^2, 119 °C, 15 min). The sterilized purees were inoculated with a lactic acid bacterium L. plantarum (10^6 CFU/g) and fermented at 36 °C for 0, 6, 8 and 10 h. The fermented product was stored at 4 °C.

2.3. Sensory Analysis of the Fermented Products

The fermented products, fermented at different times, were analyzed by 21 semi-trained judges from the Faculty of Engineering of the National University of Jujuy. Acceptability and purchase interest were studied by 9-point hedonic scale, classified from 1 to 4, meaning dislike/do not buy; 5, meaning indifferent/maybe buy; and from 6 to 9, meaning like/buy. In addition, a check all that applied the analysis (CATA) [4] was performed; attributes of flavor, aroma, texture and appearance were evaluated.

2.4. Characterization of the Final Product

The following analyses were carried out on the fermented puree: proximal composition determined in triplicate by official AOAC methods [5]; lactic acid bacteria count by serial dilution technique; plate count on MRS agar (incubation at 36 °C, 48 h) [6]; pH by direct measurement (UltraBasic digital pH meter, Denver Instrument) of the dilution of the product in water (1:10 ratio); antioxidant activity (IC50) of the methanolic extracts of the samples by scavenging the free radical DPPH [7]; color by determining the coordinates L*, a* and b* in the CIELAB system with a colorimeter (MSEZ117 HunterLab, Reston, Virginia, USA); viscosity measured with a VISCO STAR plus rotational viscometer (Fungilab Expert Series, Barcelona, Spain) at 10 rpm and 22–23 °C; EPS quantification performed by adding trichloroacetic acid (20%), refrigerated incubation, recovery with ethanol (96%) and drying [8]; and the total titratable acidity determined by AOAC 981.12 technique.

2.5. Statistical Analysis

Physicochemical results were expressed as mean value ± standard deviation. An analysis of variance (ANOVA) was carried out with a significance level of 0.05. Means were compared using Tukey's test.

CATA analysis was studied by Principal Component Analysis (PCA) to describe the samples and select the best product.

Physicochemical and sensory results were analyzed using Microsoft XLStat software 2009 (Addinsoft, Paris, France).

3. Results and Discussion

3.1. Sensory Analysis of Fermented Purees

Sensory analysis was essential for the selection of the final product since it allowed the characterization and identification of the attributes responsible for the acceptance or rejection of the judges. Acceptability of the fermented purees was greater than 50% for those at up to 8 h of fermentation; on the other hand, at 10 h of fermentation, the product presented a strong acidity due to the development of lactic acid bacteria, causing sensory rejection. The purchase intention showed the same behavior as acceptability: the product with 6 h of fermentation had the highest score, while the rest of the samples did not exceed 30% (Table 1).

Table 1. Acceptability and purchase interest of the purees.

Fermentation Time (h)	Acceptability			Purchase Interest		
	Like (%)	Dislike (%)	Indifferent (%)	Buy (%)	Do Not Buy (%)	Maybe Buy (%)
0	62	19	19	29	43	29
6	67	14	19	43	29	29
8	52	19	29	24	38	38
10	38	52	10	29	57	14

The CATA analysis was studied by Principal Component Analysis (Figure 1), and it explained 84.71% of the total variability. The samples were associated in three different groups. There was the unfermented puree with attributes such as a sweet, vinegary, strange flavor as well as a viscous texture and a nutritional appearance. There were purees at 6 and 8 h of fermentation with positive sensory attributes such as an acidic, rich, fruity flavor; a fluid texture; a pleasant, nice smell; and a good appearance. Finally, there was the puree with the longest fermentation time with nonacceptance by the judges due to the development of the bacteria, which provided undesirable characteristics such as an ugly, bitter and very acidic taste; a strange, unpleasant and intense odor; and a bad appearance. Therefore, the puree with 8 h of fermentation was selected as the best option.

3.2. Formulation of the Food and Its Characterization

The functional product obtained with the 15% *w/v* of the flours (50:50 quinoa flour: amaranth flour) with water was inoculated with 10^6 CFU/g of the bacterial strain, which is the minimum concentration required by the Argentine Food Code (2021) [9], to be considered a probiotic. After 8 h of fermentation at 36 °C, a puree with 7.60×10^8 CFU/g was obtained. The bacterial growth produced a decrease in pH and an increase in total titratable acidity. The initial pH was 4.97, a slightly acidic value, due to the presence of the strawberry pulp, reaching a final pH of 3.86; furthermore, 0.408 g of lactic acid was produced in 100 g of the sample. Table 2 shows the proximal composition of the functional product, with 100 g of the product having 19.60 g of carbohydrates, 1.74 g of proteins, 0.39 g of lipids and 89 kcal.

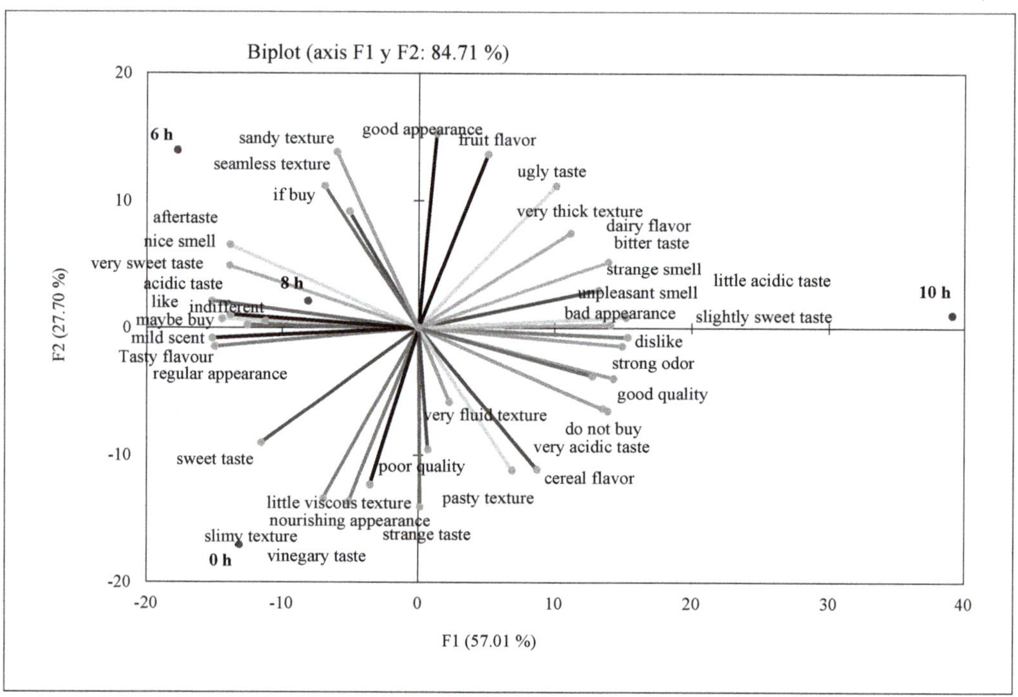

Figure 1. Principal Component Analysis of sensory attributes of purees.

Table 2. Proximal composition of the fermented puree.

	Humidity	Proteins	Lipids	Ash	Carbohydrates
Fermented puree	78.21 ± 0.10	1.74 ± 0.17	0.39 ± 0.04	0.06 ± 0.03	19.60

X ± DS = Mean ± standard deviation; n = 3 determinations.

The fermented puree had an IC50 of 10.3 mg/mL, which represents high antioxidant activity, since quinoa, amaranth and strawberry pulp provide significant amounts of compounds with antioxidant activity [10]. The fermented puree presented the following chromaticity coordinates for L*, a* and b*: 43.85 ± 1.34, 15.24 ± 0.34 and 11.72 ± 0.23, respectively. The reddish-brown color was due to the addition of strawberry pulp and the subsequent sterilization at high temperatures, which could produce non-enzymatic browning by the Maillard reaction, obtaining brown melanoidin pigments.

The viscosity of the fermented puree was 5029 mPa*s at 10 rpm, which had increased from 4126 mPa*s of the unfermented product. As the BAL used was characterized by generating EPS in the fermentation media [11], 6.78 g of EPS per liter of the fermented puree was developed during the fermentation. The production of EPS, in general, increased the viscosity and water-retention capacity, improving the rheological properties (in addition to improving the biofunctional contribution due to the soluble fiber content), but, due to the amylolytic capacity of the bacteria, it was not enough to significantly enhance the viscosity during the fermentation of the puree.

4. Conclusions

Quinoa and amaranth were excellent raw materials for the production of fermented foods, which constitutes an alternative for the revaluation of Andean crops and adds value to regional production chains. A fermented functional food based on Andean grains was elaborated with a concentration of Lactobacillus Plantarum of 10^8 CFU/g, so it could be defined as a possible probiotic food. In addition, the antioxidant characteristics of the fermented product could have a potential role in reducing the risk of contracting type 2 diabetes, cardiovascular diseases and hypertension. In this way, this food will allow for the diversification of healthy food options in the market, especially for populations with health problems associated with lactose intolerance, celiac disease and/or gluten-restricted diets as well as for the growing trend of veganism.

Author Contributions: Conceptualization, C.E.S.A.; validation, M.D.J., M.O.L. and N.C.S.; formal analysis, C.E.S.A. and M.D.J.; investigation, C.E.S.A.; resources, M.O.L. and N.C.S.; writing—original draft preparation, C.E.S.A.; writing—review and editing, M.D.J. and M.O.L.; visualization and supervision, M.O.L. and N.C.S.; project administration, M.D.J. and M.O.L.; funding acquisition, M.O.L. and N.C.S. All authors have read and agreed to the published version of the manuscript.

Funding: This work was supported by the grant Ia ValSe-Food-CYTED (119RT0567) and SECTER–Universidad Nacional de Jujuy—CONICET.

Institutional Review Board Statement: Not applicable.

Informed Consent Statement: Not applicable.

Data Availability Statement: The works consulted are detailed in the bibliography.

Conflicts of Interest: The authors declare no conflict of interest.

References

1. Corbo, M.R.; Bevilaqua, A.; Petruzzelli, L.; Casanova, F.P.; Sinigaglia, M. Functional beverages: The emerging side of functional foods. *Compr. Rev. Food Sci. Food Saf.* **2014**, *13*, 1192–1206. [CrossRef]
2. Kandylis, P.; Pissaridi, K.; Bekatoron, A.; Kanellaki, M.; Koutinas, A.A. Dairy and non-dairy probiotic beverages. *Curr. Opin. Food Sci.* **2016**, *7*, 58–63. [CrossRef]
3. Di Cagno, R.; De Angelis, M.; Limitone, A.; Minervini, F.; Carnevali, P.; Corsetti, A.; Gaenzle, M.; Ciati, R.; Gobbetti, M. Glucan and fructan production by sourdough Weissella cibaria and Lactobacillus plantarum. *J. Agric. Food Chem.* **2006**, *54*, 9873–9881. [CrossRef]
4. Pramudya, R.C.; Seo, H.-S. Using Check-All-That-Apply (CATA) method for determining product temperature dependent sensory-attribute variations: A case study of cooked rice. *Food Res. Int.* **2018**, *105*, 724–732. [CrossRef] [PubMed]
5. AOAC. *Official Methods of Analysis International*, 16th ed.; EEUU: Gaithsburgf, MA, USA; Washintong, DC, USA, 1995.
6. ICMSF. Microorganismos de los alimentos. In *Técnicas de Análisis Microbiológicos*, 2nd ed.; Editorial Acribia: Zaragoza, Spain, 1983; Volume 1, pp. 113–115.
7. Segundo, C.; Román, L.; Gómez, M.; Martínez, M. Mechanically fractionated flour isolated from green bananas (*M. cavendishii var. nanica*) as a tool to increase the dietary fiber and phytochemical bioactivity of layer and sponge cakes. *Food Chem.* **2016**, *219*, 240–248. [CrossRef]
8. Zannini, E.; Jeske, S.; Lynch, K.M.; Arendt, E.K. Development of novel quinoa based yoghurt fermented with dextran producer Weissella cibaria MG1. *Int. J. Food Microbiol.* **2018**, *268*, 19–26. [CrossRef]
9. Argentine Food Code. Chapter XVII, Article 1389. Available online: https://www.argentina.gob.ar/sites/default/files/anmat_caa_capitulo_xvii_dieteticosactualiz_2021-07.pdf (accessed on 1 January 2022).
10. Palombini, S.V.; Claus, T.; Maruyama, S.A.; Gohara, A.K.; Souza, A.H.P.; Souza, N.E.; Visentainer, J.V.; Gomes, S.T.M.; Matsushita, M. Evaluation of nutritional compounds in new amaranth and quinoa cultivars. *Food Sci. Technol.* **2013**, *33*, 339–344. [CrossRef]
11. Carrizo, S.L.; Montes de Oca, C.E.; Hébert, M.E.; Saavedra, L.; Vignolo, G.; LeBlanc, J.G.; Rollán, G. Lactic Acid Bacteria from Andean Grain Amaranth: A Source of Vitamins and Functional Value Enzymes. *J. Mol. Microbiol. Biotechnol.* **2017**, *27*, 289–298. [CrossRef]

Proceeding Paper

Antioxidant Properties of Chickpea (*Cicer arietinum* L.) Protein Hydrolysates: The In Vitro Evaluation of SOD Activity in THP-1 Cell Line †

Noelia María Rodríguez-Martin [1], José Carlos Marquez [1], Álvaro Villanueva [1], Francisco Millán [1], María del Carmen Millán-Linares [2] and Justo Pedroche [1,*]

1 Department of Food & Health, Instituto de la Grasa, CSIC. Ctra. de Utrera Km. 1, 41013 Seville, Spain
2 Department of Medical Biochemistry, Molecular Biology, and Immunology, School of Medicine, Universidad de Sevilla, Av. Dr. Fedriani 3, 41071 Seville, Spain
* Correspondence: j.pedroche@csic.es

† Presented at the IV Conference Ia ValSe-Food CYTED and VII Symposium Chia-Link, La Plata and Jujuy, Argentina, 14–18 November 2022.

Abstract: Chickpeas (*Cicer arietinum* L.) are the third most important grain legume in the world in terms of their area coverage, the volume of production and the amount of trade. The crop is primarily grown in India, being USA, Argentina and Mexico the main producers of chickpea in the American continent. It can be considered as a cheap, sustainable, and healthy source of nutrients with a high content of proteins. In addition, the hydrolysis of chickpea proteins (CPH) can release bioactive peptides with immunomodulatory properties. Hence, the in vitro study of CPH in the monocytic cell line would be a relevant strategy to recognise the immunomodulatory hydrolysates. The main aim of this study was to evaluate the immunomodulatory potential of CPH. A chickpea protein concentrate was hydrolysed using Bioprotease-660LA under specific conditions. The resulting hydrolysates were evaluated to search for the potentially bioactive CPHs. The study led to the identification of one bioactive hydrolysate, which was used on THP-1 cell line as it was stimulated with lipopolysaccharide (LPS) to evaluate the inflammatory status of it. The ELISA and RT-qPCR techniques were used to analyse the levels of inflammatory cytokine production. The total superoxide dismutase (SOD) activity was also evaluated using a commercial determination kit. Our data showed that the selected CPH down-regulated the mRNA transcriptional levels of cytokines IL-1β and TNF-α in the LPS-stimulated THP-1 cell line. In addition, CPH increased the SOD activity, in contrast to the LPS control. This study suggests that CPH may improve the inflammatory state and play a significant role in the regulation of the SOD signalling pathways in THP-1 cell lines.

Keywords: chickpea; immunomodulatory; protein hydrolysate; superoxide dismutase; THP-1; inflammation

Citation: Rodríguez-Martin, N.M.; Marquez, J.C.; Villanueva, Á.; Millán, F.; Millán-Linares, M.d.C.; Pedroche, J. Antioxidant Properties of Chickpea (*Cicer arietinum* L.) Protein Hydrolysates: The In Vitro Evaluation of SOD Activity in THP-1 Cell Line. *Biol. Life Sci. Forum* **2022**, *17*, 12. https://doi.org/10.3390/blsf2022017012

Academic Editors: Norma Sammán, Mabel Cristina Tomás, Loreto Muñoz and Claudia Monika Haros

Published: 27 October 2022

1. Introduction

Chickpeas (*Cicer arietinum* L.) are the third most important grain legume in the world in terms of their area coverage, the volume of production and the amount of trade [1]. The crop is primarily grown in India, being USA, Argentina and Mexico the main producers of chickpea in the American continent. [1]. It can be considered a cheaper, sustainable, and healthy source of nutrients with a high content of immunonutrients. Chickpea varieties contain 15–40% proteins, 15–68% carbohydrates, and 7% lipids. Nutritionally, this legume has a good composition of the essential amino acids with a high bioavailability, and a low allergenicity when it is compared to others such as soybeans or lupins. Furthermore, chickpea protein hydrolysates (CPHs) have better technological properties than chickpea concentrates and isolates do, in addition to the possibility of them producing bioactive peptides [2,3].

Inflammation is a normal process that removes the injurious infection and promotes the tissue repair and recovery related to several inflammatory pathways and involving Reactive Oxygen Species (ROS) [4–9]. However, the presence of certain factors has been linked to the inhibition of the inflammation resolution, thereby causing a chronical low-grade inflammation. This chronical inflammatory status causes several alterations to the normal cellular physiology, thus increasing the risk of non-communicable disease, infectious susceptibility, and cancer development risks [10]. There are multiple possible causes for the development of a chronical inflammatory status: bad diet habits, intestinal dysbiosis, obesity, stress, and a high level of physical activity. The clinical consequences of this chronical inflammatory status are associated with a higher risk of metabolic syndrome and its related comorbidities, cardiovascular disease, cancer, neurodegenerative disorders, autoimmune diseases, among others [10].

The study of CPH in the THP-1 cell line is a relevant strategy to recognise the antioxidant and anti-inflammatory immunonutrient in the treatment of inflammatory diseases.

2. Materials and Methods

2.1. Material and Reagent

Chickpea protein hydrolysate (CPH) was supplied by the Group of Plant Proteins at the Instituto de la Grasa-CSIC (Spain). THP-1 human cell line was supplied by the Unit of Cell Biology of the Instituto de la Grasa-CSIC (Spain). All of the solvents and buffers that were used were of biomolecular grade, and all of the disposable plastics were DNase, RNase-free and sterile.

2.2. Cell Culture and Treatment

THP-1 cell line was cultured in RPMI-1640, 10% FBS, and it was 1% penicillin-streptomycin-supplemented. After seeding THP-1 cells into the 12 well plates at concentration of 500,000 cell/mL, the cells were stimulated with 0.1 μg/mL LPS with the exception of the negative control. The stimulated cells were treated with CPH at different doses (50 and 100 μg/mL), except for the negative (no LPS, No CPH) and positive control (LPS; no CPH), and they were incubated for 24 h (37 °C, 5% CO_2).

2.3. Cell Cytotoxicity MTT Assay

THP-1 cell line was seeded in a 96 well plate at concentration of 500,000 cell/mL. After seeding, the cells were subjected to different concentrations of CPH (0, 25, 50, 100, 250, 500 and 1000 μg/mL) and incubated for 24 h. Afterwards, MTT solution was applied at these concentrations to each well and incubated for 3 h. Then, DMSO was applied to dissolve the formazan crystals. The absorbance was obtained using a microplate reader at 570 nm and it was corrected to 650 nm. The cell cytotoxicity of CPH at different concentrations was compared to the negative (non-treated cells) and positive controls (Triton).

2.4. Gene Expression Analysis and Cytokine Measurement

Real-time quantitative polymerase chain reaction (qRT-PCR) was used to measure the levels of the messenger RNA (mRNA). The cells were recovered, and the total RNA was extracted using TRIsure reagent. Complementary DNA (cDNA) were obtained using the iScript cDNA Synthesis Kit from the extracted RNA samples. The performed primer pairs and SYBR green master mix were used to carry out RT-PCR. Finally, the TNF-α, IL-1β, SOD1, and SOD2 gene expression were normalised using the internal control GADPH (Table 1).

After the treatments were performed, the supernatants of cells containing only culture media were collected. TNF-α and IL-1β concentrations were measured using human ELISA kits following the manufacturer's instructions (Diaclone, Besancon, France).

Table 1. Primer sequences for real time PCR.

Gene Name	NM Code	Forward Sequence	Reverse Sequence
IL-1β	NM_000576.2	5′-TCCTTCAGACACCCTCAACC-3′	5′-AGGCCCCAGTTTGAATTCTT-3′
TNF-α	NM_000594.3	5′-CTGTCCTGCGTGTTGAAAGA-3′	5′-TTCTGCTTGAGAGGTGCTGA-3′
SOD1	NM_000454.5	5′-ACAAAGATGGTGTGGCCGAT-3′	5′-AACGACTTCCAGCGTTTCCT-3′
SOD2	NM_000636.4	5′-AACAACCTGAACGTCACCGA-3′	5′-CACGTTTGATGGCTTCCAGC-3′
GADPH	NM_002046.6	5′-GAGTCAACGGATTTGGTCGT-3′	5′-GACAAGCTTCCCGTTCTCAG-3′

Abbreviations: IL (Interleukine), TNF (tumor Necrosis Factor), SOD (Super Oxide Dismutase), SOD1 (Mn like SOD or extracellular SOD), SOD2 (Cu/Zn like SOD or mitochondrial SOD), GADPH (Glyceraldehyde-3-Phosphate Dehydrogenase).

2.5. Nitric Oxide Measurement

The total NO levels were obtained from the supernatants by the Griess reaction that is based on the reduction of nitrates to nitrites. Briefly, 100 μL of Griess reagent was added to 100 μL of sample. The Griess reagent consist of a mixture of equal volumes of N-(1-naphthyl) ethylenediamine (0.1%) in deionised water and sulfanilamide in 5% H_3PO_4 (1%). Absorbance at 546 nm was measured using a microplate reader. The nitrite concentration was calculated using a $NaNO_2$ standard curve.

2.6. Superoxide Dismutase Assay

After the seeding procedure, the LPS-activated THP-1 cells in a 96-well plate, and concentrations of 50 and 100 μg/mL of CPH were used and incubated for 24 h. The total activity of antioxidant enzyme SOD was measured in cellular homogenates using commercial colorimetric assay. The assay procedure was prepared according to the manufacturer's protocol (Abcam, Boston, MA, USA).

2.7. Statistical Analysis

Data are expressed as arithmetic means with standard deviations (SD). Graph Pad Prism Version 8.0.1 software (San Diego, CA, USA) was used to evaluate the data. One-way ANOVA following Tukey's test for multiple comparison and post hoc test were used to calculate the statistical significance of differences. P-value less than 0.05 was considered statistically significant. All of the assays were performed in quadruplicate.

3. Results and Discussion

Chickpeas have been shown to have further biological activities by different active compounds such as peptides, phenolic compounds, and others. Several chickpea-derivate products have been identified as compounds with activities such as antioxidant, antihypertensive, hypocholesterolemia, and anticancer [3]. In addition, the bioactive peptides, protein hydrolysates, and protein extracts from the chickpeas have also showed the properties that have been mentioned before [2–4]. In this sense, the biological activity of a CPH might be related to the group of peptides that were obtained during the enzyme hydrolysis where the amino acid composition, conformation, length, and sequence are key factors that produce the biological activity.

3.1. CPH Diminishing Inflammation in LPS Stimulated THP-1 Cells

The cell viability that was evaluated by using the MTT method was not affected by any of the CPH concentrations that were used. It may, therefore, be concluded that the use of CPH could be of interest in human nutrition as a non-toxic active ingredient. Thus, 50, 100, 250 and 500 μg/mL CPH were selected for the experiments (Figure 1).

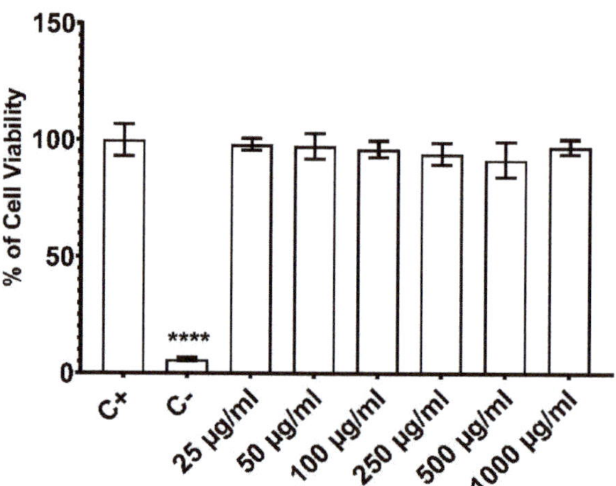

Figure 1. Effect of CPH on THP-1 cells viability. THP-1 cells were cultured with different doses of CPH (25, 50, 100, 250, 500 and 1000 µg/mL) for 24 h. Data are expressed as mean ± SD of four separate experiments. p value < 0.0001 was considered statistical significance (****). C+, live cell control; C− death cell control.

To evaluate the immunomodulatory properties of CPH, the gene expression of the proinflammatory cytokines was measured by RT-qPCR in the THP-1 cells that were stimulated with LPS after 24 h of exposure. The CPH treatment reduced the statistically significant ($p < 0.05$) expression of *TNF-α* by 20% at a dose of 50 µg/mL, as shown in Figure 2A. This effect was particularly important in the case of IL-1β, where the *IL-1β* mRNA expression decreased by 58%, 48%, 62% and 64% with 50, 100, 250 and 500 µg/mL CPH, respectively ($p < 0.0001$) in comparison that which was observed in the LPS-treated cells (Figure 2B). Furthermore, the cytokine release that was measured by the ELISA method, which is shown in Figure 2D, was consistent with these RT-qPCR results. The major effect of CPH in IL-1β suggests the inhibition of the inflammatory pathways that are involved in the release of IL-1β. In the case of the LPS stimulation, the proinflammatory status has been produced by the binding of it with Toll Like Receptor 4 (TLR4). When the TLR4 receptor signalling pathway is activated, the aNF-κB protein complex is assembled, and it acts as transcriptional factor inside the nucleus and promotes the expression of the pro-inflammatory genes such as *Pro-Il-1β*, *Pro-Il-18*, *Pro-Caspase-1*, *Asc*, and *Nlrp3*. The expression of these proteins leads to the inflammasome pathway activation and the assembly of NLRP3, ASC, and Pro-Caspase-1, thereby conforming the NLRP3 inflammasome. Caspase-1 is released from the inflammasome complex, and this promotes the activation of the Il-1β and Il-18 cytokines from their precursors which are secreted to the extracellular medium [6]. Hence, the CPH could act as a targeting drug by inhibiting the Il-1β expression and the release of the NF-κB transcriptional effect activation, the inflammasome assembly or a mix of both pathways. However, these pathways are too long, and it is necessary that we conduct more research to verify the inflammatory markers and pathway that are mentioned above.

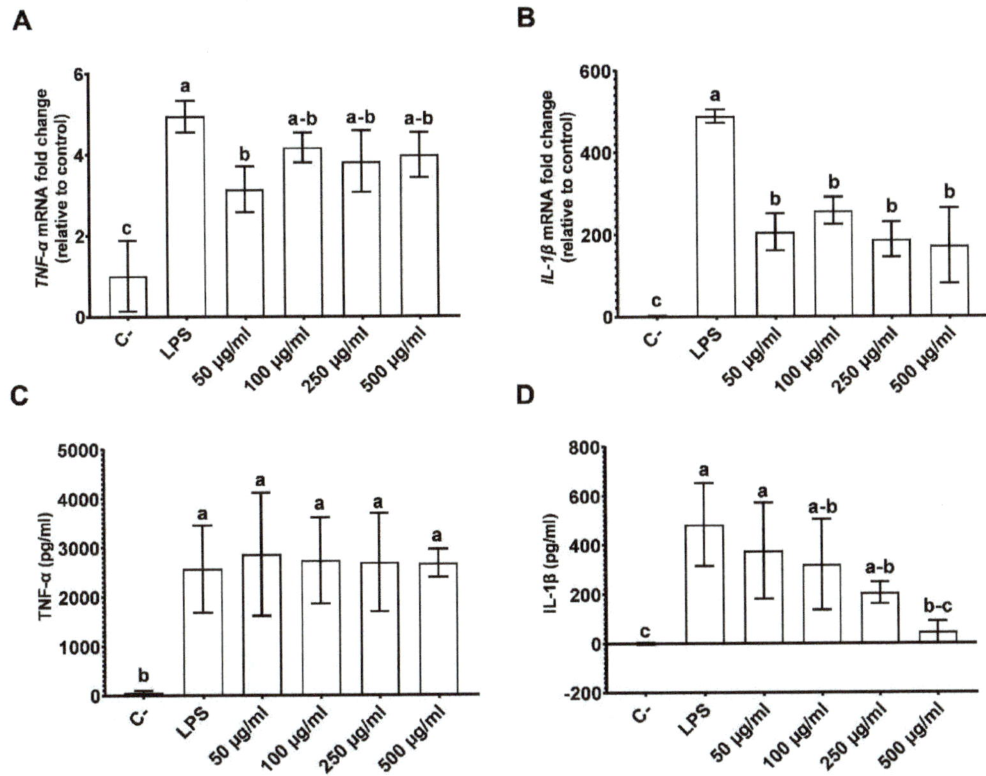

Figure 2. Effect of CPH on LPS-induced cytokines mRNA relative expression (*TNF-α* expression were significantly lower in the C−, and 50 µg/mL (**A**), and *IL-1β* which shown a significant expression decrease in all the concentrations as well as C− (**B**)) in THP-1 cells, and effect of CPH on LPS-induced protein level (which did not show statistical significance in the TNF-α release for any concentration of product (**C**), in contrast to the release of IL-1β which were shown a significant decrease in a dose dependant manner (**D**)). THP-1 cells previously stimulated with LPS were cultured with different doses of CPH (50, 100, 250 and 500 µg/mL) for 24 h. Data are expressed as mean ± SD of four separate experiments. p value < 0.05 was considered statistical significance and marked with different letters. C−, live cell control; LPS, lipopolysaccharide-treated cell.

3.2. The Effects of CPH on Oxidative Stress

To investigate the antioxidant cellular responses to CPH, the reactive nitrogen species levels in the supernatants (NO), the total SOD inhibition rate in the cellar homogenates and the extracellular SOD (SOD1) and the mitochondrial SOD (SOD2) transcriptional levels were determined. The results revealed that CPH influences the NO levels which revealed a dose-dependent increase in the amount of extracellular NO (Figure 3A). Likewise, the SOD activity showed a recovery of the SOD levels ($p < 0.05$) (Figure 3B). Finally, the mRNA of the SOD isoforms revealed that the effect of the *SOD1* mRNA expression was similar to the negative control (Figure 3C). In the case of the levels of *SOD2*, a decrease in the mRNA expression of this isoform was observed which accounts for about 55% of it ($p < 0.0001$), 64% of it ($p < 0.0001$), 56% of it ($p < 0.0001$), and 72% of it ($p < 0.0005$), respectively, for each dose in contrast to the LPS control (Figure 3D).

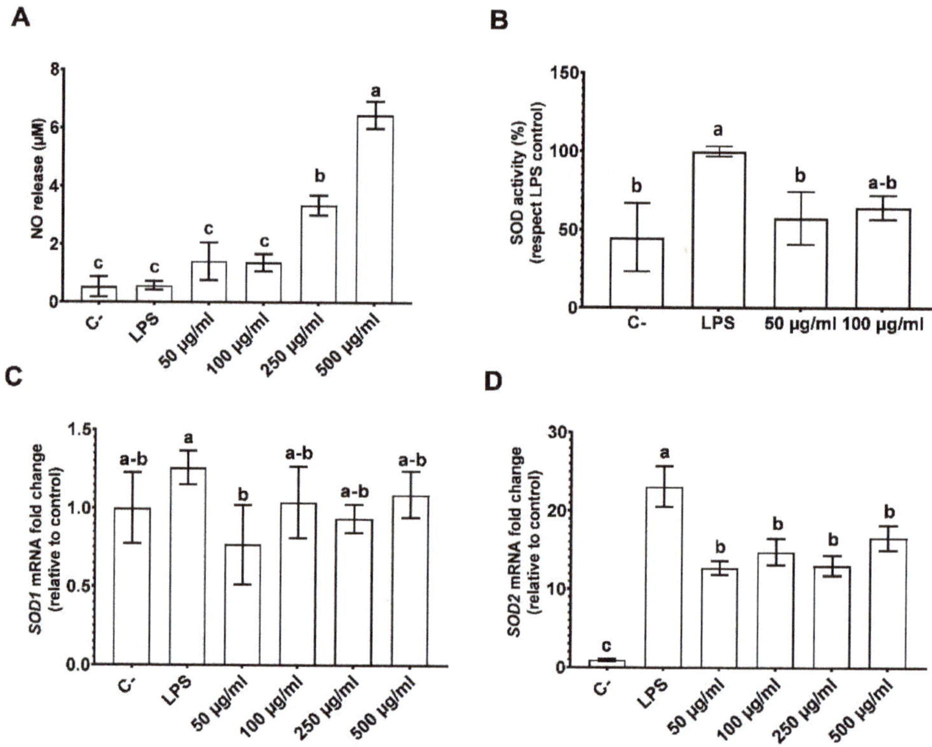

Figure 3. The effect of CPH was high in a concentration dependant manner on induced NO release (**A**), a restore of the SOD activity was observed in contrast to the LPS control (**B**), and LPS-induced ROX enzymes mRNA relative expression, *SOD1* (**C**) and a lower *SOD2* expression was observed when the cells were treated with CPH at different doses (**D**) in THP-1 cells. THP-1 cells previously stimulated with LPS were cultured with different doses of CPH (50, 100, 250 and 500 µg/mL) for 24 h (in exception of SOD activity assay). Data are expressed as mean ± SD of four separate experiments. p value < 0.05 was considered statistical significance and marked with different letters. C− live cell control; LPS, lipopolysaccharide-treated cell.

NO, NO_2, and $ONOO^-$ are reactive nitrogen/oxygen species which are considered both as signalling molecules and potentially cytotoxic ones. The superoxide molecules react with NO, forming ONOO, and this reaction is catalysed by the SODs, specially SOD1 in the cytoplasm. The ONOO molecule is considered as a strong oxidant and nitrating agent, but it is also is stable at physiological conditions [7]. The intracellular ROS, specifically the mitochondrial ROS are new actors enhancing the proinflammatory pathways including the NF-κB and NLRP3 inflammasome. The cellular SOD activity that is shown in Figure 3B suggests that there was a decrease of the SOD levels which could be indicated as a restoration of intracellular ROS basal levels. Similar results appeared in the *SOD1* and *SOD2* transcriptional levels which may be traduced in an intracellular ROS and mitochondrial ROS stress decrease.

Besides this, NO·is considered to be a widespread signalling molecule in mammals. NO has been shown to be a cell-to-cell messenger, being responsible for the modulating processes that are related to the circulation as vasodilation and relaxation of the smooth muscles [8]. SOD enzymes also participate in cell signalling pathways, regulating NO release [8,9]. As it was observed in Figure 3A, the NO mRNA expression with the negative and positive controls showed similar values to those of the NO mRNA expression with 50 and 100 µg/mL of the CPH doses. In addition, in the tolerance studies of low-dose LPS in the THP-1 cells, it has been reported that the NO levels increase, and the down-regulating of cytokines occurs [9]. This tolerance to LPS is associated with the epigenetic changes in the chromatin [9]. The CPH could generate the epigenetic changes that mimic to LPS low-dose-tolerance, thereby affecting the NO production and cytokine transcription.

4. Conclusions

Our data showed that the CPH down-regulated the mRNA transcriptional levels of the proinflammatory cytokines in THP-1 cell line that was stimulated with LPS, and particularly, the levels of IL-1β may indicate the NLRP3 inflammasome inhibition by the repression of the NF-κB inflammatory pathway. In addition, CPH has restored the intracellular SOD activity, in contrast to that which was achieved by the LPS control, mainly by the use of the mitochondrial SOD which removed the ROS stress and prevented the NF-κB and NLRP3 inflammasome feeding. Finally, the increase of the NO release at the highest CPH doses could indicate a possible cell-to-cell signal to restore the non-oxidative and non-inflammatory cellular status. Herein, this study suggests that CPH may improve the inflammatory states, and that it has a role in the SOD-derived oxidative stress in the THP-1 cell line.

Author Contributions: Conceptualization, J.P. and N.M.R.-M.; methodology, N.M.R.-M., and J.C.M.; software, J.C.M.; validation, J.P., M.d.C.M.-L. and Á.V.; formal analysis, N.M.R.-M.; investigation, J.P.; resources, J.P.; data curation, N.M.R.-M.; writing—original draft preparation, N.M.R.-M.; writing—review and editing, J.P.; visualization, J.C.M.; supervision, F.M.; project administration, J.P.; funding acquisition, J.P., and F.M. All authors have read and agreed to the published version of the manuscript.

Funding: This work was supported by grant Ia ValSe-Food-CYTED (119RT0567), by the "Centro para el Desarrollo Tecnológico Industrial" (CDTI), project no: IDI-20200562 (Diseño y evaluacion de nuevos ingredientes aplicados a alimentacion deportiva basados en hidrolizados proteicos vegetales (PLANTPROSPORT)), as well as the "Ministerio de Ciencia, Innovación y Universidades", and the "Agencia Estatal de Investigación", project no: PID2019-111368RB-I00 (Evaluation of new disruptive technologies (steam explosion) in the design of tailor-made plant protein hydrolysates applied to sport nutrition (PROVERDE)).

Institutional Review Board Statement: Not applicable.

Informed Consent Statement: Not applicable.

Data Availability Statement: Not applicable.

Conflicts of Interest: The authors declare no conflict of interest.

References

1. FAOSTAT. Available online: https://www.fao.org/faostat/en (accessed on 1 August 2022).
2. Boukid, F. Chickpea (*Cicer arietinum* L.) protein as a prospective plant-based ingredient: A review. *Int. J. Food Sci. Technol.* **2021**, *56*, 5435–5444. [CrossRef]
3. Faridy, J.-C.M.; Stephanie, C.-G.M.; Gabriela, M.-M.O.; Cristian, J.-M. Biological Activities of Chickpea in Human Health (*Cicer arietinum* L.). A Review. *Plant Foods Hum. Nutr.* **2020**, *75*, 142–153. [CrossRef] [PubMed]
4. Guilliams, M.; Mildner, A.; Yona, S. Developmental and Functional Heterogeneity of Monocytes. *Immunity* **2018**, *49*, 595–613. [CrossRef]
5. Liu, T.; Zhang, L.; Joo, D.; Sun, S.-C. NF-κB signaling in inflammation. *Signal Transduct. Target. Ther.* **2017**, *2*, 17023. [CrossRef]
6. Swanson, K.V.; Deng, M.; Ting, J.P.Y. The NLRP3 inflammasome: Molecular activation and regulation to therapeutics. *Nat. Rev. Immunol.* **2019**, *19*, 477–489. [CrossRef] [PubMed]
7. Duez, H.; Pourcet, B. Nuclear Receptors in the Control of the NLRP3 Inflammasome Pathway. *Front. Endocrinol.* **2021**, *12*, 630536. [CrossRef] [PubMed]
8. Wang, Y.; Branicky, R.; Noë, A.; Hekimi, S. Superoxide dismutases: Dual roles in controlling ROS damage and regulating ROS signaling. *J. Cell Biol.* **2018**, *217*, 1915–1928. [CrossRef] [PubMed]
9. Rios, E.C.S.; de Lima, T.M.; Moretti, A.I.S.; Soriano, F.G. The role of nitric oxide in the epigenetic regulation of THP-1 induced by lipopolysaccharide. *Life Sci.* **2016**, *147*, 110–116. [CrossRef]
10. Furman, D.; Campisi, J.; Verdin, E.; Carrera-Bastos, P.; Targ, S.; Franceschi, C.; Ferrucci, L.; Gilroy, D.W.; Fasano, A.; Miller, G.W.; et al. Chronic inflammation in the etiology of disease across the life span. *Nat. Med.* **2019**, *25*, 1822–1832. [CrossRef]

Proceeding Paper

Study of the Residue from *Salvia hispanica* (Chia) Seed Oil Extraction by Cold Pressing for Repurposing as Functional Food to Prevent Metabolic Syndrome †

Gabriela Alarcon [1,2], Agostina Valoy [1], Analia Rossi [1] and Susana Jerez [1,2,*]

1 Instituto Superior de Investigaciones Biológicas (INSIBIO), Consejo Nacional de Investigaciones Científicas y Técnicas (CONICET), Universidad Nacional de Tucumán (UNT), Av Independencia 1800 y Chacabuco 461, San Miguel de Tucumán, Tucumán 4000, Argentina

2 Facultad de Ciencias Naturales, Universidad Nacional de Tucumán (UNT), Miguel Lillo 205, San Miguel de Tucumán, Tucumán 4000, Argentina

* Correspondence: sjerez@herrera.unt.edu.ar

† Presented at the IV Conference Ia ValSe-Food CYTED and VII Symposium Chia-Link, La Plata and Jujuy, Argentina, 14–18 November 2022.

Abstract: Despite its high nutritional value, the residue obtained by cold pressing of the chia seeds (expeller) is still undervalued. Expeller is rich in proteins and fibers and contains about 7% omega-3 α-linolenic acid (ALA). Considering that chia seed has been reported to improve insulin resistance, among other cardiovascular risk factors, the aim of this work was to study the effects of a diet enriched with the expeller on a rabbit model of metabolic syndrome. A nutritional analysis of the expeller was evaluated. Rabbits were fed a standard diet (CD), a high-fat diet (HFD) and a HFD in which 20% of the calories from fat were replaced by the expeller (ED). At the end of the 6-week feeding period, clinical, biochemical, and vascular reactivity studies were performed. Results: The ED did not modify body weight or visceral fat, and reached to control fasting glucose (mg/dL; CD: 113 ± 3; HFD: 1261 ± 5; ED: 90 ± 7), insulin resistance (area under the curve of glucose tolerance, CD: 612 ± 23; HFD: 676 ± 17; ED: 517 ± 38), triglycerides (mg/dL CD: 113 ± 14; HFD: 192 ± 22; ED = 98 ± 22) and the TyG index (CD: 8.3 ± 0.2; HFD: 9.3 ± 0.3; ED: 8.28 ± 0.23). With respect to the vascular studies, a blunted norepinephrine response was found. In conclusion, the results showed a promising use of the expeller to develop functional foods that improve metabolic syndrome symptoms.

Keywords: expeller; chia; insulin resistance; functional food; metabolic syndrome; residue

Citation: Alarcon, G.; Valoy, A.; Rossi, A.; Jerez, S. Study of the Residue from *Salvia hispanica* (Chia) Seed Oil Extraction by Cold Pressing for Repurposing as Functional Food to Prevent Metabolic Syndrome. *Biol. Life Sci. Forum* **2022**, *17*, 13. https://doi.org/10.3390/blsf2022017013

Academic Editors: Norma Sammán, Mabel Cristina Tomás, Loreto Muñoz and Claudia Monika Haros

Published: 27 October 2022

1. Introduction

Recycling and reuse are actions that allow the circularity of the system, with tasks that reintroduce residues as new inputs in the production circuits, adopting a circular production model [1]. The residue from chia seed (expeller) is a by-product obtained from the oil extraction of seeds by cold pressing [2]. The expeller has a high nutritional value due to its high content of proteins, fibers and minerals. In addition, the expeller has a residual content of 7–11% rich in omega-3 alpha-linolenic acid (ALA) oil [3]. However, this by-product is undervalued: it is often used for animal feed or marketed as crushed chia. To this date, some studies for adding value to expeller have been reported: the production of bioactive peptides with antioxidant properties by enzymatic hydrolysis with papain [2] and the development of gluten-free premixes with buckwheat and chia flour for a bread product [4]. In the last years, the demand for functional foods has increased in developed economies as people look for a safer way to improve their general health and living. Functional foods are designed to meet the provision of basic nutrients as well as to offer the potential to enhance health or reduce the risk of diseases.

Metabolic syndrome (MS) identifies a subgroup of individuals with a shared pathophysiology who are at high risk of developing cardiovascular disease (CVD), a leading cause of death in the world [5]. In previous work, we have characterized a model of MS by feeding rabbits on a high-fat diet (HFD) for 6 weeks. This model shows normal body weight, hypertriglyceridemia, a reduction in the high density lipoprotein cholesterol (HDL-C), an increase in visceral abdominal fat (VAF), fasting glucose (FG) and glucose intolerance. Furthermore, a HFD causes an early vascular dysfunction involving a proinflammatory status [6]. The beneficial effects of chia seeds on CVD risk factors have been widely studied. However, we recently reported that, in our experimental model of MS, the partial replacement of corn oil with chia oil has a partially beneficial effect on MS [7]. Thus, considering that expeller is an undervalued residue, despite its high nutritional value, and chia seed has been reported to improve insulin resistance, among other CVD risk factors, the aim of this work was to study the effects of a rich-in-expeller diet on a rabbit model of MS, validating its use as a functional food.

2. Materials and Methods

2.1. Chia Expeller

The expeller was provided by NYNAGRO, a local chia seed and oil producer through its brand CHIA VITA (Tucumán, Argentina).

2.2. Animal Handling and Diets

The experimental protocols for this study were approved by the Institutional Animal Care and Use Committee from the National University of Tucuman. All animal care and use programs were performed according to the *Guide for the Care and Use of Laboratory Animals* (National Institute of Health Publication, 8th edition, updated 2011). Eighteen male hybrid rabbits, 45 days after weaning, were individually housed in metal cages in a room under controlled conditions. The animals were randomized, separated, and fed with different diets for 6 weeks: regular rabbit chow (n = 6, control diet, CD); a CD supplemented with 18% fat (n = 6, high-fat diet, HFD), and a HFD in which 20% of the calories from fat were replaced by the expeller (n = 6, expeller diet, ED).

2.3. Preparation of Rich-in-Expeller Diet

The diet was prepared by replacing 10% of the daily calories ingested by the animal. An amount of expeller that replaced 10% of the total calories was incorporated into the balanced feed for rabbit (Ganave, Buenos Aires, Argentina). In summary, 100 g of rich-in-expeller food was prepared by adding 11.4 g of expeller (360 cal/100 g) into 84.8 g of the food matrix (balanced feed with 10% corn oil plus 8% lard added, 412 cal/100 g). Thus, a HFD and ED were isocaloric.

2.4. Proximate Composition of Expeller and Rich-in-Expeller Diet

The proximate composition was determined according to the Association of Official Analytical Chemists (AOAC). The crude fat was determined by continuous extraction in a Soxhlet extraction system (using an IVA glass extractor, Buenos Aires, Argentina), during 8 h using petroleum ether (fraction 60–80) as a solvent at 60 °C (AOAC 922.06).

The determination of ashes, was carried out in a muffle furnace brand ORL-Hornos Electricos, Lomas de Zamora, Buenos Aires, Argentina, model ORL-IV, at 550 °C (AOAC 923.03).

The crude protein content was determined by the Kjeldahl method (AOAC 984.13), using for digestion and subsequent distillation a Buchi K-435 equipment (Buchi Labortechnik, Flawil, Switzerland). To transform the nitrogen content obtained into protein, factor 6.25 was used. The moisture content was obtained by drying in an oven at 65 °C to a constant weight (AOAC 925.09).

2.5. Clinical and Biochemical Parameters

An intraperitoneal glucose tolerance test was performed two days before the end of the 6 weeks of feeding. At the end of the dietary intervention, food was withdrawn for 12 h. The rabbits were anesthetized, and the mean arterial blood pressure (MAP) was measured [6]. Blood samples were collected through the catheter inserted into the carotid artery. Plasma total cholesterol (TC), HDL-C, low density lipoprotein cholesterol (LDL-C), triglycerides (TG) and FG were measured by using colorimetric reactions with commercial kits (Wiener, Rosario, Argentina). A midline incision was made in the rabbit using surgical techniques, and the adipose tissues from the abdominal areas were collected and weighed. The VAF and TyG indices were calculated [6,7].

2.6. Vascular Function Assessment

The aorta was carefully dissected, and the isometric contractile response was measured [6]. The endothelial function was checked by building a concentration–response curve (CRC) to acetylcholine (Ach, 10^{-8}–10^{-5} M) in phenylephrine-precontracted aortic rings. Contractile responses to the vasoconstrictor agonists, angiotensin II (Ang II) and norepinephrine (NE), were checked. One group of arteries was exposed to increasing doses of Ang II (10^{-10}–10^{-6} M), and the other group was exposed to increasing doses of NE (10^{-8}–10^{-4} M) to construct a CRC. The KCl contractile response was also checked.

2.7. Statistical Analyses

The CRCs were fitted using a nonlinear interactive fitting program (GraphPad Prism 3.0; GraphPad Software Inc., San Diego, CA, USA). The agonist potencies were calculated as pEC_{50} (negative logarithm of the molar concentration of an agonist producing 50% of the maximum response), and the maximum response was expressed as the Rmax (the maximum effect elicited by the agonist). The investigators were blinded to the treatment until the data analysis. The results are reported as a mean ± standard error of the mean (SEM). The Shapiro and Wilk goodness-of-fit test was used to test for normal distribution. Statistically significant differences were calculated by the one-way analysis of variance (followed by Duncan's post-test); $p < 0.05$ was considered statistically significant.

3. Results

3.1. Composition of Expeller and Rich-in-Expeller Diet

Table 1 shows the proximate composition of the expeller and ED.

Table 1. Proximate composition of chia expeller and rich-in-expeller diet.

	Expeller	ED
Moisture	4.5 ± 0.3	5.8 ± 0.6
Lipid	9.1 ± 1.1	7.6 ± 2.1
Protein	29.9 ± 0.06	14.1 ± 0.4
Ash	5.2 ± 0.24	7.82 ± 0.18

3.2. Clinical and Biochemical Parameters

At the end of the feeding trials (6 weeks), the animal body weights did not differ significantly among the groups (Table 2). The MAP was lower in the rabbits fed an ED than that in the other diet groups (Table 2). The FG, TyG index (Table 2) and area under the curve from the intraperitoneal glucose tolerance test (AUCglu, nmol/Lxmin) were significantly lower in the ED than those of the HFD group (CD: 612 ± 23 vs. HFD = 676 ± 17 vs. ED = 517 ± 38; $n = 6$, $p < 0.05$, one-way ANOVA and Duncan's post-test). Regarding the lipid profile, the ED significantly reduced the serum levels of TG and LDL-C with respect to the HFD. However, the ED did not modify VAF with respect to the HFD.

Table 2. Clinical and biochemical parameters.

	CD	HFD	ED
Body Weight (g)	1940 ± 149	2043 ± 46	2059 ± 79
Visceral abdominal fat (%)	1.09 ± 0.17	2.31 ± 0.14 [a]	2.85 ± 0.22 [a]
Fasting Glucose (mg/dL)	113.0 ± 3.0	126.1 ± 5.8 [a]	90.0 ± 7.7 [b]
Total Cholesterol (mg/dL)	59.0 ± 6.0	77.7 ± 4.8 [a]	54 ± 8 [b]
LDL-cholesterol (mg/dL)	29.0 ± 7.8	39.8 ± 7.2	15.47 ± 3.6 [b]
HDL-cholesterol (mg/dL)	51.5 ± 6.9	24.2 ± 2.8 [a]	22 ± 4.8 [a,b]
Triacylglycerol (mg/dL)	105 ± 14	192 ± 22 [a]	94 ± 15 [b]
TyG index	8.3 ± 0.2	9.3 ± 0.3 [a]	8.3 ± 0.2 [b]
Blood pressure (mmHg)	56.0 ± 2.6	56.7 ± 5.3	40.0 ± 6.3 [a,b]
Heart rate (bpm)	265 ± 25	281.8 ± 27	306 ± 47

Data are expressed as mean ± SEM of 6 rabbits fed either a control diet (CD), high-fat diet (HFD) or expeller diet (ED). [a] $p < 0.05$ indicates statistically significant differences between HFD or ED groups compared to rabbits fed on CD. [b] $p < 0.05$ indicates statistically significant differences between rabbits fed on HFD and rabbits fed on ED (one-way ANOVA and Duncan's post-test).

3.3. Effects of Diets on Acetylcholine Relaxation Responses

Ach-relaxation was significantly lower in the arteries from the HFD and ED groups than that in the arteries from the CD group (CD-Rmax: 60 ± 4% vs. HFD-Rmax: 43 ± 5% vs. ED: 49 ± 2%; $n = 13$; $p < 0.05$, one-way ANOVA and Duncan's post-test).

3.4. Effect of Diets on Contractile Response to Agonists

The Rmax to Ang II was similar in all diet groups. However, pEC_{50} was higher in the arteries from the EDs than that of the arteries from HFDs (CD: 7.82 ± 0.08 vs. HFD: 8.13 ± 0.06 vs. ED = 8.25 ± 0.09, $n = 8$, $p < 0.05$; one-way ANOVA and Duncan's post-test). Thus, a shift to the left of the curve was found (Figure 1).

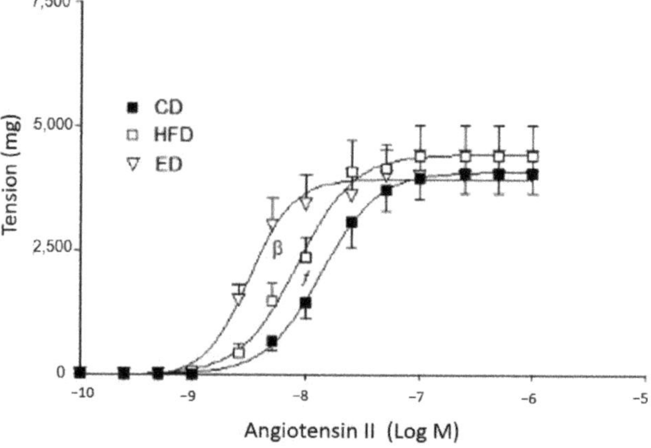

Figure 1. Concentration–response curve to angiotensin II. CD: regular diet; HFD: high-fat diet; ED: expeller diet. [β] $p < 0.05$ indicates statistically significant differences between pEC_{50} of arteries from rabbits fed on ED and the other diet groups (one-way ANOVA). [f] $p < 0.05$ indicates statistically significant differences between pEC_{50} of arteries from rabbits fed on HFD and CD (one-way ANOVA).

The Rmax to NE was significantly decreased in the arteries from the ED group as compared to the other diet groups (Figure 2). The contractile response to KCl was higher in the arteries from HFD than that of CD. The ED did not normalize the contractile response to KCl (CD: 3907 ± 270 vs. HFD: 6743 ± 500 mg vs. ED: 5668 ± 695 mg; $n = 7$; $p < 0.05$, one-way ANOVA and Duncan's post-test).

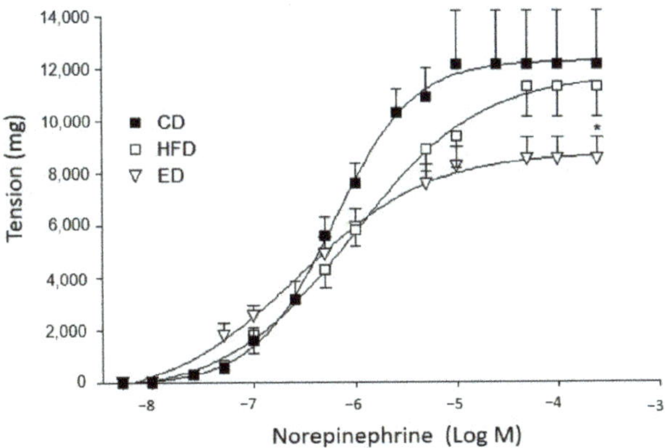

Figure 2. Concentration–response curve to norepinephrine. CD: regular diet; HFD: high-fat diet; ED: expeller diet. * $p < 0.05$ indicates statistically significant differences between Rmax of arteries from animals fed on ED and the other diet groups (one-way ANOVA).

4. Discussion

In previous work, we demonstrated that the replacement of 3% corn oil with chia oil into HFD reduces the TG levels and improves endothelial dysfunction [7]. The increase in plasmatic ALA levels may account for these effects. Data from the literature report the beneficial effect of chia seeds on insulin resistance [8]. Unexpectedly, chia oil worsened FG and insulin resistance in our model of MS. Dietary fiber is strongly related to the prevention of chronic diseases, such as hypercholesterolemia, diabetes and obesity. Taking into account that a high fiber content has been found in chia expeller [3], we hypothesized that an ED might improve FG and insulin resistance in our model. The results from rabbits fed on an ED showed that TG reached the control levels and reduced FG, insulin resistance and the TyG index. These parameters are key CVD risk factors characterizing MS. In addition, the MAP was decreased in rabbits fed on an ED. Chia seed hydrolysates have been found to inhibit ACE [9]. This may be the mechanism explaining the reduction in MAP. However, while the expeller protein content was high, the protein content in the ED was similar to that of the CD [6]. This means further studies on protein composition should be carried out to clarify this point. Furthermore, ACE inhibition induces the upregulation of the Ang II receptors (AT1) as a compensatory mechanism. As the in vitro isometric contractions evaluate Ang II-AT1 receptor binding pharmacologically, we hypothesized that AT1 receptor upregulation might account for the increasing Ang II sensitivity found in the arteries of rabbits fed on an ED. Regarding the contractile response to NA, ED reduced the Rmax with respect to the other groups. Considering that insulin blunts sympathetic vasoconstriction [10], this effect of an ED may be related to the improvement of insulin sensitivity.

5. Conclusions

The results showed promising use of the expeller to develop functional foods that improve metabolic syndrome symptoms.

Author Contributions: S.J.: Conceptualization, Formal analysis, Resources, Funding Acquisition, Writing-Reviewing and Editing; G.A.: Conceptualization, Methodology, Formal analysis, Writing; A.V.: Methodology, Investigation, Formal Analysis. Analia Rossi: Methodology, Investigation. All authors have read and agreed to the published version of the manuscript.

Funding: This research was supported by: PICT 2015/1164 (ANCyT-Argentina); PIUNT G621 (SCAIT, Argentina) and by institutional funds from Instituto Superior de Investigaciones Biológicas (INSIBIO). Furthermore, present work was supported by grant Ia ValSe-Food-CYTED (119RT0567).

Institutional Review Board Statement: All experimental protocols were approved by the Institutional Animal Care and Use Committee from the National University of Tucuman, Argentina, Approval number: 021/2019; Date: 15 November 2019; Date endorsed: 23 March 2021) according to the Guide for the Care and Use of Laboratory Animals (National Institute of health Publication 8th edition, updated 2011).

Informed Consent Statement: Not applicable.

Data Availability Statement: The data that support the findings of this study are available from the corresponding author, Susana Jerez, upon reasonable request.

Acknowledgments: The authors thank Agustina Gonzalez Colombres for technical support.

Conflicts of Interest: The authors declare no conflict of interest.

References

1. Instituto Nacional de Tecnología Industrial. *El Valor de los Residuos: Distintos Modos de Reducir, Reutilizar, Reciclar y Revalorizar Residuos Industriales*, 1st ed.; Instituto Nacional de Tecnología Industrial: Buenos Aires, Argentine, 2016. Available online: https://www.inti.gob.ar/publicaciones/descargac/1 (accessed on 20 May 2022).
2. Cotabarren, J.; Rosso, A.; Tellechea, M.; García-Pardo, J.; Rivera, J.L.; Obregón, W.D.; Parisi, M.G. Adding value to the chia (*Salvia hispanica* L.) expeller: Production of bioactive peptides with antioxidant properties by enzymatic hydrolysis with Papain. *Food Chem.* **2019**, *274*, 848–856. [CrossRef] [PubMed]
3. Capitani, M.; Spotorno, V.; Nolasco, S.; Tomas, M. Physicochemical and functional characterization of by-products from chia (*Salvia hispanica* L.) seeds of Argentina. *LWT-Food Sci. Technol.* **2012**, *45*, 94–102. [CrossRef]
4. Coronel, E.; Guiotto, E.; Aspiroz, M.; Tomas, M.; Nolasco, S.; Capitani, M. Development of gluten-free premixes with buckwheat and chia flours: Application in a bread product. *LWT-Food Sci. Technol.* **2021**, *141*, 110916. [CrossRef]
5. Fahed, G.; Aoun, L.; Zerdan, B.M.; Allam, S.; Bou Zerdan, M.; Bouferraa, Y.; Assi, H.I. Metabolic Syndrome: Updates on Pathophysiology and Management in 2021. *Int. J. Mol. Sci.* **2022**, *23*, 786. [CrossRef]
6. Alarcón, G.; Roco, J.; Medina, A.; Sierra, L.; Medina, M.; Jerez, S. High fat diet-induced metabolically obese and normal weight rabbit model shows early vascular dysfunction: Mechanisms involved. *Int. J. Obes.* **2018**, *42*, 1535–1543. [CrossRef] [PubMed]
7. Alarcón, G.; Medina, A.; Martin Alzogaray, F.; Medina, M.; Roco, J.; Jerez, S. Partial replacement of corn oil with chia oil into a high fat diet produces either beneficial and deleterious effects on metabolic and vascular alterations in rabbits. *PharmaNutrition* **2020**, *14*, 100218. [CrossRef]
8. Chicco, A.G.; D'Alessandro, M.E.; Hein, G.J.; Oliva, M.E.; Lombardo, Y.B. Dietary chia seed (*Salvia hispanica* L.) rich in alpha-linolenic acid improves adiposity and normalises hypertriacylglycerolaemia and insulin resistance in dyslipaemic rats. *Br. J. Nutr.* **2009**, *101*, 41–50. [CrossRef] [PubMed]
9. San Pablo-Osorio, B.; Mojica, L.; Urías-Silvas, J.E. Chia Seed (*Salvia hispanica* L.) Pepsin Hydrolysates Inhibit Angiotensin-Converting Enzyme by Interacting with its Catalytic Site. *J. Food Sci.* **2019**, *84*, 1170–1179. [CrossRef] [PubMed]
10. Limberg, J.K.; Soares, R.N.; Power, G.; Harper, J.L.; Smith, J.A.; Shariffi, B.; Jacob, D.W.; Manrique-Acevedo, C.; Padilla, J. Hyperinsulinemia blunts sympathetic vasoconstriction: A possible role of β-adrenergic activation. *Am. J. Physiol. Regul. Integr. Comp. Physiol.* **2021**, *320*, R771–R779. [CrossRef] [PubMed]

Proceeding Paper

Chia Oil-in-Water Nanoemulsions Produced by Microfluidization †

Nicolás Censi, Luciana M. Julio and Mabel C. Tomás *

Centro de Investigación y Desarrollo en Criotecnología de Alimentos (CIDCA), Consejo Nacional de Investigaciones Científicas y Técnicas (CONICET), Comisión de Investigaciones Científicas de la Provincia de Buenos Aires (CICPBA), Universidad Nacional de La Plata (UNLP), 47 y 116, La Plata 1900, Buenos Aires, Argentina

* Correspondence: mabtom@hotmail.com

† Presented at the IV Conference Ia ValSe-Food CYTED and VII Symposium Chia-Link, La Plata and Jujuy, Argentina, 14–18 November 2022.

Abstract: Oil-in-water (O/W) nanoemulsions (d < 200 nm) are systems with considerable potential for protecting and delivering sensible ingredients such as chia seed oil rich in ω-3 fatty acids (~64% α-linolenic acid). These systems can be formed by applying either low- or high-energy methods. High-pressure homogenization, microfluidization and sonication are included within the latter. The main aim of this research work was to obtain and characterize chia oil-in-water nanoemulsions by microfluidization. Therefore, O/W nanoemulsions with 10% (*w*/*w*) chia oil and 2% (*w*/*w*) sodium caseinate were prepared at three levels of microfluidization pressure: 600, 1000 and, 1200 bar. Droplet sizes of the nanoemulsions expressed as the Sauter mean diameter, were found between 108 to 125 nm. Additionally, the resulting superficial droplet charge was between −37 to −41 mV. The global stability of the different systems was evaluated through the evolution of their backscattering for 50 days. In this sense, nanoemulsions obtained at 1000 and 1200 bar recorded high global stability, while those obtained at 600 bar showed some signs of destabilization. In terms of oxidative stability, all systems studied recorded low values of primary and secondary oxidation products as a function of storage, as determined by peroxide value index (PV) and thiobarbituric acid reactive substances (TBARs) assays, respectively. The omega-3 fatty acid content of the nanosystems was also evaluated, without significant changes during the storage period. Thus, chia O/W nanoemulsions obtained by microfluidization proved to be suitable delivery systems for bioactive compounds of chia seed, with potential applications in the development of functional food.

Keywords: chia oil; delivery systems; microfluidization; omega-3; O/W emulsion

Citation: Censi, N.; Julio, L.M.; Tomás, M.C. Chia Oil-in-Water Nanoemulsions Produced by Microfluidization. *Biol. Life Sci. Forum* **2022**, *17*, 14. https://doi.org/10.3390/blsf2022017014

Academic Editors: Norma Sammán, Loreto Muñoz and Claudia Monika Haros

Published: 28 October 2022

1. Introduction

The lower content of saturated fatty acids, the adequate concentration of linoleic acid (18–20%), and the high content of alpha-linolenic acid (55–60%) make chia oil an appealing option for healthy food and cosmetic applications [1]. Consumption of sufficiently high levels of ω-3 fatty acids has been linked to reduced risk of certain chronic diseases, such as inflammation, cardiovascular disease, immune response, and mental disorders [2]. However, due to the high content of unsaturated fatty acids, chia oil is susceptible to lipid oxidation and needs to be protected for its potential incorporation into foods. In this sense, emulsion-based delivery systems have proved to be an effective strategy to protect and deliver sensitive ingredients. Nanoemulsions can be used as delivery systems due to their extended stability and improved bioavailability [3]. In these systems, the smaller sizes of droplets give them more stability against gravitational separation and droplet aggregation than conventional emulsions [4]. Among the different approaches available for forming nanoemulsions, microfluidization has been shown to be efficient at producing small droplets with uniform particle size distributions. During the microfluidization process, a combination of high disruptive forces (cavitation, turbulence, and shear) leads to the formation of very fine oil droplets [5].

In this context, the aim of this work was to study the incorporation of chia oil into functional O/W nanoemulsions and to evaluate the effect of the homogenizer pressure level on its physicochemical properties.

2. Materials and Methods

2.1. Materials

Commercial chia oil extracted by cold pressing was purchased by SDA S.A. (Lobos, Argentina) and stored at 4 ± 1 °C until use in an amber bottle. Sodium caseinate (NaCas) was purchased from Sigma Chemical Company (St. Louis, MO, USA). All reagents used were of analytical grade.

2.2. Preparation of O/W Emulsions

Chia oil-in-water (O/W) nanoemulsions with an oil:water ratio of 10:90, and 2% NaCas as an emulsifying agent, were obtained. In the first homogenization step, the aqueous phase and the chia oil were mixed using an Ultraturrax T-25 (Janke & Kunkel GmbH, Staufen, Germany) at 13,500 rpm for 2 min. A second step was carried out using a Microfluidizer (LM10, Microfluidics, Westwood, CA, USA) by applying three different pressures (600, 1000 and 1200 bar) and three passes. Nisin and potassium sorbate were added to all systems to prevent microbial growth. The O/W nanoemulsions were stored at 4 °C for 50 days protected from light.

2.3. Emulsions Characterization

2.3.1. Droplet Size

Droplet size was determined by laser diffraction with a particle size analyzer (Mastersize 2000E, Malvern Instrument Ltd., Worcestershire, UK). For each measurement, approximately 1 mL of sample was diluted directly in the water bath of the dispersion system with a pump speed of 1700 rpm (Hydro 2000G), reaching an obscuration of 10–12%. Measurements were performed in sextuplicate.

2.3.2. Emulsion Stability

The physical stability of the emulsions was monitored by periodic measurements of dispersed light using a Vertical Scan Analyzer Quick Scan (Coulter Corp., Miami, FL, USA) according to Pan et al. (2002) [6].

2.3.3. ζ-Potential

The ζ-potential was determined at room temperature using a Zeta Potential Analyzer (Brookhaven 90Plus/Bi-MAS, New York, NY, USA) according to Julio et al. (2015) [7]. For each determination, 0.05 g of the sample was dispersed in 100 mL of milli-Q water. Measurements were performed in duplicate.

2.3.4. Rheological Properties

Rheological measurements were performed in an AR-G2 stress-controlled oscillatory rheometer (TA instrument, New Castle, DE, USA) at 25 ± 0.3 °C with a cone-plate sensor system. Samples were subjected to a shear rate increase (1 to 500 s^{-1}) for 3 min, maintained at 500 s^{-1} for 1 min, and a decrease in shear rate (500 to 1 s^{-1}) for 3 min.

2.4. Oxidative Stability of Emulsions

To evaluate the oxidative stability of chia O/W nanoemulsions during the storage period, primary and secondary oxidative products were determined through peroxide value (PV) and 2-thiobarbituric acid reactive substances (TBARs), respectively. The PV was carried out according to the spectrophotometric method described by Diaz et al. (2003) [8], while the TBARs assay was according to Hu and Zhong (2010) [9]. In addition, chia oil was extracted from the different nanoemulsions and its fatty acid profile was obtained using Agilent Technologies 7890A equipment at the initial time and then 50 days of storage.

2.5. Statically Analysis

Analysis of variance (ANOVA) was performed on the data set using Statgraphics Centurion XV (StatPoint Technologies, Warrenton, VA, USA). II. Fisher's test (LSD) was also applied at a significance level of 5%.

3. Results and Discussion

3.1. Emulsion Characterization

The droplet size distribution (DSD) of chia O/W emulsions obtained at 600, 1000 and 1200 bar presented similar bimodal curves (data not shown) with a major peak between ~0.040 to 0.500 µm and a small shoulder ranging from ~0.500 to 1.010 µm. In addition, it was observed that droplet size distribution curves were narrower at higher pressure levels, denoting a higher degree of uniformity for Cas1000 and Cas1200 systems; the droplet sizes ($D_{3.2}$) of the systems are presented in Table 1. It could be observed that there was a significant influence ($p \leq 0.05$) of the microfluidization pressure on the droplet size, resulting in the following increasing order Cas1000 < Cas1200 < Cas600. Although a decrease in particle size would be expected as a function of pressure increase [10], this did not occur for samples Cas1000 and Cas1200. This could be because the emulsifying agent was not sufficient to fully stabilize the droplet interface, favoring the formation of larger particles during homogenization.

Table 1. De Sauter mean diameter ($D_{3.2}$) and ζ-potential of chia O/W nanoemulsions.

System	ζ-Potential (mV)	$D_{3.2}$ (µm)	
		0 d	50 d
Cas600	-41 ± 4^{a}	0.127 ± 0.001^{aF}	0.124 ± 0.001^{aG}
Cas1000	-40 ± 1^{a}	0.107 ± 0.02^{bH}	0.109 ± 0.002^{bH}
Cas1200	-35 ± 4^{b}	0.125 ± 0.00^{cI}	0.1240 ± 0.0005^{aJ}

Different letters indicate significant differences in the means ($p \leq 0.05$). Lowercase: differences in the same column. Uppercase: differences in the same line, for storage times of 0 and 50 days.

Regarding ζ-potential, all systems recorded negatively charged oil droplets ranging from −35 to −41 mV (Table 1). Cas1200 recorded the smallest ($p \leq 0.05$) surface charge between them, possibly because of the high-pressure treatment, which could have resulted in the unfolding of the protein structure affecting their ionic and hydrophobic interactions (Xinfa Ma et al. 2021) [11].

The global stability of the emulsions was studied through the evolution of their backscattering profiles (%BS) during storage (Figure 1). As can be seen, Cas1000 and Cas1200 presented high global stabilities, maintaining their BS profiles without significant changes throughout the whole period studied. The emulsion prepared at 600 bar exhibited slight signs of destabilization by creaming, showing a simultaneous decrease of the %BS at the bottom of the tube and an increase of this parameter at the top of it.

The rheological data from emulsions measurements were fitted to the power-law model obtaining the K (consistency coefficient) and n (flow behavior index) parameters. All systems recorded values of n~1, denoting Newtonian behavior. Besides, K values resulted between 1.77–1.64 $\times 10^{-3}$ Pa.s^{n}. In this sense, C600 samples had lower ($p \leq 0.05$) K values, which could be associated with their larger droplet size.

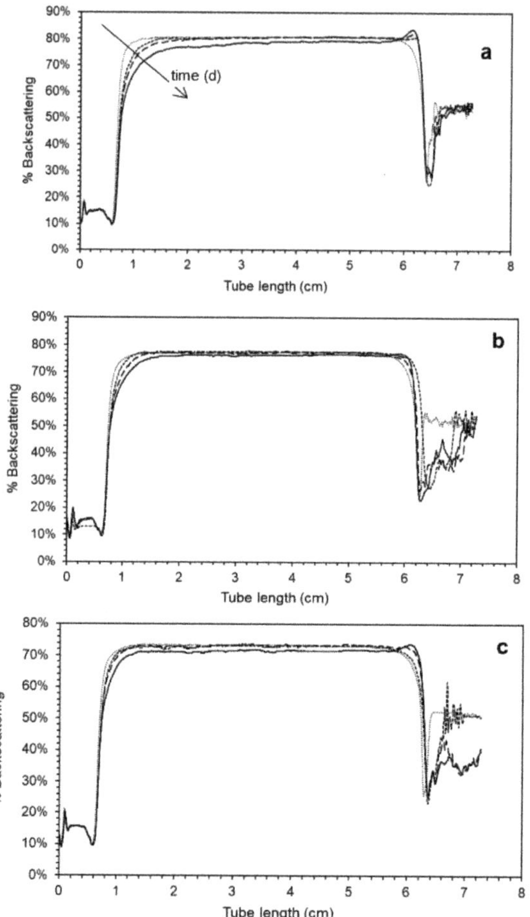

Figure 1. Backscattering patterns of chia O/W emulsions prepared at three levels of homogenization pressure (**a**) 600 (**b**) 1000 and (**c**) 1200 bar as a function of time (....) 1 d, (----) 21 d, (— —) 30 d, and (——) 50 d of storage. Average values ($n = 2$).

3.2. Oxidative Stability of Emulsions

The evaluation of the primary lipid oxidation was carried out through emulsions PV assays during the storage period. Initially, all emulsions recorded low PV values indicating that the applied emulsification process did not affect the emulsified chia oil. After storage, the primary oxidation products of all systems remained low without exceeding the maximum limit of 10 meq of hydroperoxides/kg oil established by the Codex Alimentarius (Figure 2). Regarding the secondary oxidation monitored through TBARS measurements, the three systems studied presented low values of this parameter. In addition, the omega-3 fatty acid content of the nanosystems was also evaluated, without significant changes during the storage period. Based on these results, chia O/W nanoemulsions exhibited higher oxidative stability after 50 days of refrigerated storage.

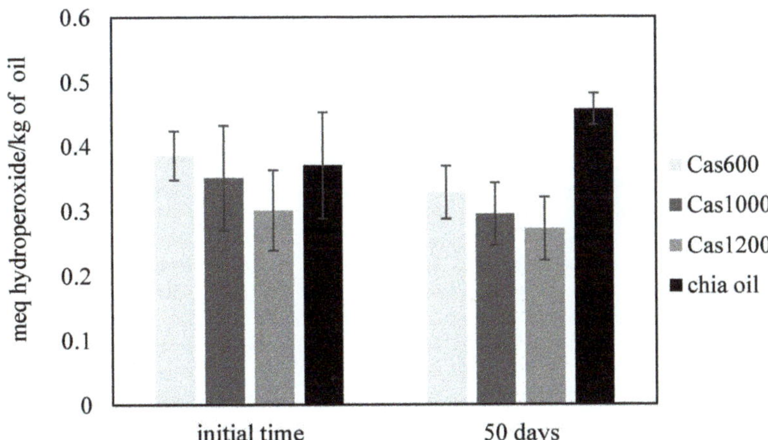

Figure 2. Peroxide value (meq of hydroperoxide/kg of oil) of chia O/W nanoemulsions at the initial time and after 50 days of refrigerated storage.

4. Conclusions

From the concentration of the emulsifying agent and the conditions of the microfluidization applied in the present study, it was possible to obtain stable chia O/W emulsions with nanoscale droplet sizes.

It was found that microfluidization pressure had a significant effect on the properties of chia O/W nanoemulsions. In this sense, emulsions obtained at 1000 and 1200 bar recorded smaller droplet sizes and higher global stability than those prepared at 600 bar. Regarding the oxidative stability of the nanoemulsions, all systems recorded low values of primary and secondary oxidation products after 50 days at 4 °C. Thus, chia nanoemulsions obtained by microfluidization technology proved to be suitable delivery systems that can protect chia seed oil.

Author Contributions: N.C.: Conceptualization, Methodology, Investigation, Formal analysis, Writing—original draft preparation; L.M.J.: Conceptualization, Methodology, Investigation, Formal analysis, Writing—review & editing, Supervision, Project administration; M.C.T.: Conceptualization, Methodology, Writing—review & editing, Supervision, Project administration, Funding acquisition. All authors have read and agreed to the published version of the manuscript.

Funding: This work was supported by grants La ValSe-Food-CYTED (119RT0567) Spain, Universidad Nacional de La Plata (X907), Agencia Nacional de Promoción Científica y Tecnológica (PICT 2019-01775; 2020-01274), CONICET PIP 2007, Argentina.

Institutional Review Board Statement: Not applicable.

Informed Consent Statement: Not applicable.

Data Availability Statement: Not applicable.

Acknowledgments: Authors wish to thank Mariela Fernandez (CETMIC) and Mariana Pennisi (CIDCA) for their technical assistance.

Conflicts of Interest: The authors declare no conflict of interest.

References

1. Imran, M.; Nadeem, M.; Manzoor, M.F.; Javed, A.; Ali, Z.; Akhtar, M.N.; Muhammad, A.; Yasir, H. Fatty acids characterization, oxidative perspectives and consumer acceptability of oil extracted from pre-treated chia (*Salvia hispanica* L.) seeds. *Lipids Health Dis.* **2016**, *15*, 1–13. [CrossRef] [PubMed]
2. Tur, J.A.; Bibiloni, M.M.; Sureda, A.; Pons, A. Dietary sources of omega 3 fatty acids: Public health risks and benefits. *Br. J. Nutr.* **2012**, *107*, S23–S52. [CrossRef]

3. Bai, L.; McClements, D.J. Development of microfluidization methods for efficient production of concentrated nanoemulsions: Comparison of single-and dualchannel microfluidizers. *J. Colloid Interf. Sci.* **2016**, *466*, 206–212. [CrossRef] [PubMed]
4. Komaiko, J.S.; Mcclements, D.J. Formation of food-grade nanoemulsions using low-energy preparation methods: A review of available methods. *Compr. Rev. Food Sci. Food Saf.* **2016**, *16*, 331–352. [CrossRef] [PubMed]
5. Monroy-Rodríguez, I.; Monroy-Villagrana, A.; Cornejo-Mazon, M.; García-Pinilla, S.; Hernandez-Sánchez, H.; Gutiérrez-Lopez, G.F. Microfluidization in nanofood engineering. In *Food Engineering Series*; Hebbar, U., Ranjan, S., Dasgupta, N., Kumar Mishra, R., Eds.; Springer International Publishing: Cham, Switzerland, 2020; pp. 153–175. [CrossRef]
6. Pan, L.G.; Tomás, M.C.; Añón, M.C. Effect of sunflower lecithins on the stability of water-in-oil and oil-in-water emulsions. *J. Surfactants Deterg.* **2002**, *5*, 135–143. [CrossRef]
7. Julio, L.M.; Ixtaina, V.Y.; Fernández, M.A.; Torres Sánchez, R.M.; Wagner, J.R.; Nolasco, S.M.; Tomás, M.C. Chia seed oil-in-water emulsions as potential delivery systems of ω-3 fatty acids. *J. Food Eng.* **2015**, *162*, 48–55. [CrossRef]
8. Díaz, M.; Dunn, C.M.; McClements, D.J.; Decker, E.A. Use of caseinophosphopeptides as natural antioxidants in oil-in-water emulsions. *J. Agric. Food Chem.* **2003**, *51*, 2365–2370. [CrossRef] [PubMed]
9. Hu, Z.; Zhong, Q. Determination of thiobarbituric acid reactive substances in microencapsulated products. *Food Chem.* **2010**, *123*, 794–799. [CrossRef]
10. McClements, D.J. Emulsion design to improve the delivery of functional lipophilic components. *Annu. Rev. Food Sci. Technol.* **2010**, *1*, 241–269. [CrossRef] [PubMed]
11. Ma, X.; Chatterton, D.E. Strategies to improve the physical stability of sodium caseinate stabilized emulsions: A literature review. *Food Hydrocoll.* **2021**, *119*, 1–14. [CrossRef]

Proceeding Paper

Substitution of Critical Ingredients of Cookie Products to Increase Nutritional Value †

Giorgos Myrisis [1,2], Silvia Aja [1] and Claudia Monika Haros [1,*]

1 Instituto de Agroquímica y Tecnología de alimentos, Consejo Superior de Investigaciones Cieníficas (IATA-CSIC), 46980 Valencia, Spain

2 School of Food and Nutritional Sciences, Agricultural University of Athens, 118 55 Athens, Greece

* Correspondence: cmharos@iata-csic.es

† Presented at the IV Conference Ia ValSe-Food CYTED and VII Symposium Chia-Link, La Plata and Jujuy, Argentina, 14–18 November 2022.

Abstract: In accordance with the current technological advances in the bakery industry, to increase the nutritional value of cookies without affecting their technological and sensorial parameters, critical ingredients of cookie products (flour, sugar, butter) were substituted with whole quinoa flour and by-products of chia (oil and fibre). To optimize the formulation of the cookie (% substitution of critical ingredients) and the baking conditions, factorial design and the response surface methodology were applied. The optimal formulation presented significantly higher amounts of protein content with an improved amino acid profile and ash and fibre contents, while caloric values decreased compared to the control sample. Concerning baking conditions, 170–180 °C and 11 min were found to be the most appropriate conditions for the control formulation. The same conditions were then applied to the optimized formula. As cereal products are one the sources of acrylamide and as many studies have indicated its carcinogen potential, its concentration in the cookie products was investigated. The International Agency of Research on Cancer classified acrylamide as a probable human carcinogen (Group 2A). The results indicated that the acrylamide levels were lower than the limit concentrations of the regulation (EU) 2017/2158 (350 μg/kg). The results showed that all cookies have acceptable technological and sensorial quality, and the new cookies have highly nutritious properties and health benefits.

Keywords: chia by-products; cookie; nutritional value; quinoa flour; acrylamide

Citation: Myrisis, G.; Aja, S.; Haros, C.M. Substitution of Critical Ingredients of Cookie Products to Increase Nutritional Value. *Biol. Life Sci. Forum* **2022**, *17*, 15. https://doi.org/10.3390/blsf2022017015

Academic Editors: Norma Sammán, Mabel Tomás and Loreto Muñoz

Published: 28 October 2022

1. Introduction

Nowadays, the relationship between nutrition and health has become a trend in constant evolution because consumers are aware of the importance of their diet, so they consume foods that are low in fat, salt, and/or calories and high in nutritional value.

Baked products are an important part of one's diet, and today, a great variety of such products can be found by adding innovative ingredients to improve their nutritional characteristics [1]. In particular, cookies are frequently eaten by all segments of the population [2].

Quinoa (*Chenopodium quinoa*) and chia (*Salvia hispanica* L.) are ingredients that stand out for their high nutritional value, containing essential fatty acids (omega-3 and omega-6), high mineral and vitamin content [3], high amounts of fibre [1], and high biological-value proteins and essential amino acids [4]. Therefore, adding these ingredients to the formulation of cookies will nutritionally improve their properties.

As cereal products are a source of acrylamide, and as many studies have indicated its carcinogen potential, its concentration in the cookie products was investigated. The International Agency of Research on Cancer has classified acrylamide as a probable human carcinogen (Group 2A).

This study aimed to investigate the utility of chia and quinoa products as ingredients to produce cookies with high nutritional value and to evaluate their technological and sensorial quality.

2. Materials and Methods

2.1. Materials

The raw materials were commercial flour, whole royal quinoa flour (*Chenopodium quinoa* Willd), and fibre chia fraction (*Salvia hispanica* L.) from BENEXIA Company (Santiago, Chile). Commercial butter, sunflower oil, chia oil, and nonfat dry milk were purchased from local market (Valencia, Spain).

2.2. Cookie Production

The cookie control dough formula was made according to the Approved American Association of Cereal Chemists (AACC) method 10-52 [5]. The cookie formulations with quinoa, chia fibre, and/or chia/sunflower oil were obtained by replacement flour, sugar, and butter, respectively, at three levels (0, 50, and 100% substitution following a factorial design 3^3 (Table 1).

Table 1. Ingredients substituted.

1. Critical Ingredients	2. Substitute
3. Vegetable fat or butter	4. Sunflower/Chia Oil
5. Wheat Flour	6. Whole Quinoa flour
7. Sugar	8. Chia fibre *

* Included Stevia.

2.3. Baking Conditions

The conditions for making the cookies were optimized (170–180 °C and 11 min) and were used to elaborate the control, optimized, and 100% substituted formulations.

2.4. Proximate Composition of the Cookies

Analyses of the cookies were completed to determine the total dietary fibre (TDF) and starch according to the approved AOAC Methods 991.43 and 996.11, respectively [6]. Protein determination was performed with the Dumas combustion method and a nitrogen conversion factor of 6.25 (ISO/TS 16634-2) [7], whereas the lipids and ash contents were determined according to the AACC Methods 30-10 and 08-03 [8], respectively.

2.5. Preliminary Sensory Evaluation

The preliminary sensory evaluation was carried out by 20 untrained testers who consume cookies in their everyday life. The parameters that were evaluated were: aspect, texture, taste, and overall acceptability on a 9-point hedonic scale (9) "Especially like"; (8) "Very much like"; (7) "Moderately like"; (6) "Somewhat like"; (5) "Neither like nor dislike"; (4) "Slightly dislike"; (3) "Moderately dislike"; (2) "Very much dislike"; and (1) "Especially dislike" [9].

2.6. Determination of Acrylamide

The UNE-EN 16618:2015 method was implemented to determine acrylamide concentration in the cookie samples (Eurofins Ecosur Laboratory, Murcia, Spain).

2.7. Statistical Analysis

To establish significant differences between samples ($p < 0.05$), one-way ANOVA and Fisher's least significant differences (LSD) were applied.

3. Results

The optimal formulation, taking into account technological and nutritional parameters, was observed at around 60% substitution of wheat flour for whole quinoa flour, around 80% substitution of butter for chia-sunflower oil mix, and around 50% substitution of sugar for chia fibre fraction with stevia.

To corroborate that the optimal cookie formulation had a better nutritional value than the control, a comparison between samples was completed. The substitution with whole quinoa flour or chia fibre increased the mineral content; the substituted sample at 100% had the highest values, followed by the optimized sample and the control sample. The results are shown in Figure 1. Protein values and total dietary fibre values showed the same pattern as for ashes. The results obtained were due to quinoa having a high content of total minerals (ash). The caloric values of the three samples are represented in Figure 2, where the control formulation has the highest value, followed by the optimized sample, and the 100% substituted formulation showed the lowest value. These results were explained by the fact that the control sample has a significant quantity of simple carbohydrates and is poor in dietary fibre.

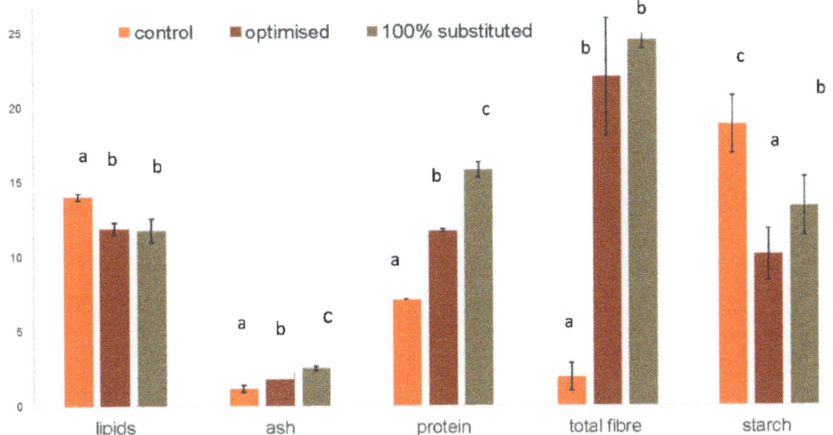

Figure 1. Proximate composition of cookies. Data are expressed as g/100 g in dry matter, mean ± standard deviation ($n = 3$). Values in bars followed by the same letter are not significantly different at 95% confidence level ($p < 0.05$).

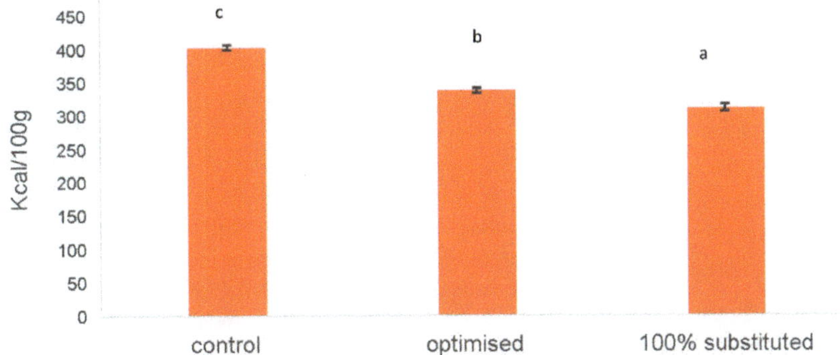

Figure 2. Caloric values of cookies. Data are expressed as mean ± standard deviation ($n = 3$). Values in bars followed by the same letter are not significantly different at 95% confidence level ($p < 0.05$).

According to the preliminary sensory evaluation, all the studied parameters showed the same pattern: the high value corresponds to the control, followed by the optimized sample and the lowest values for the 100% substituted cookie (Figure 3). However, the samples are acceptable by consumers, with overall acceptability for the optimal formulation between 7 and 8 on the hedonic scale.

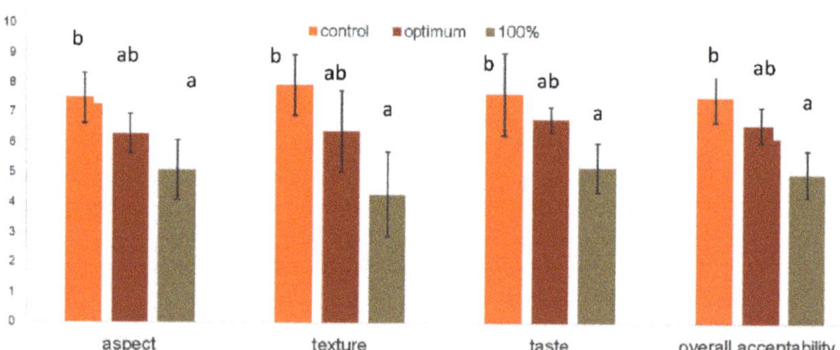

Figure 3. Preliminary sensory analysis on hedonic scale of nine points. Data are expressed as mean ± standard deviation (n = 3). Values in bars followed by the same letter are not significantly different at 95% confidence level ($p < 0.05$).

The International Agency of Research on Cancer [10] has classified acrylamide as a possible human carcinogen (Group 2A). In Figure 4, we can observe the values obtained for acrylamide in the three studied formulations. The results indicate that the highest acrylamide levels were found in the control samples, followed by the 100% substituted cookies, with the lowest value in the optimized sample. All the formulations presented lower values than the recommended limit of the EU Regulation 2017/2158, lower than 350 µg/kg for biscuits and wafers [11].

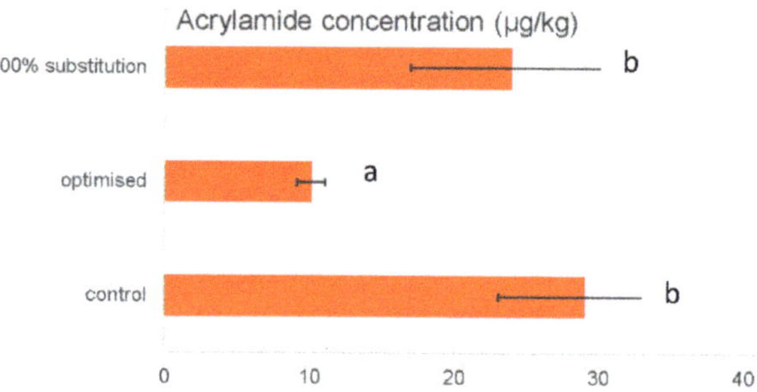

Figure 4. Acrylamide content in cookies. Data are expressed as mean ± standard deviation (n = 3). Values in bars followed by the same letter are not significantly different at 95% confidence level ($p < 0.05$).

4. Conclusions

The optimized sample is highly nutritious and has a similar technological quality to the control sample with lower harmful components, such as acrylamide, than the control. Therefore, the optimized cookie, with the replacement of critical ingredients, is a nutritional and healthy alternative to common cookies for a better diet.

Author Contributions: Conceptualization, C.M.H.; methodology, G.M. and S.A.; validation, G.M. and S.A.; formal analysis, G.M.; investigation, G.M., S.A. and C.M.H.; resources, C.M.H.; writing—original draft preparation, G.M.; writing—review and editing, C.M.H.; visualization, G.M. and

C.M.H.; supervision, C.M.H.; project administration, C.M.H.; funding acquisition, C.M.H. All authors have read and agreed to the published version of the manuscript.

Funding: This work was supported by grant Ia ValSe-Food-CYTED (119RT0567) and Food4ImNut (PID2019-107650RB-C21) from the Ministry of Sciences and Innovation (MICINN-Spain).

Institutional Review Board Statement: Not applicable.

Informed Consent Statement: Not applicable.

Acknowledgments: Giorgos Myrisis's fellowship from the Erasmus Program of the Agricultural University of Athens is acknowledged.

Conflicts of Interest: The authors declare no conflict of interest.

References

1. Alcântara Brandão, N.; Borges de Lima Dutra, M.; Andrade Gaspardi, A.L.; Segura Campos, M.R. Chia (*Salvia hispanica* L.) Cookies: Physicochemical/Microbiological Attributes, Nutrimental Value and Sensory Analysis. *Food Meas.* **2019**, *13*, 1100–1110. [CrossRef]
2. Brito, I.L.; de Souza, E.L.; Felex, S.S.S.; Madruga, M.S.; Yamashita, F.; Magnani, M. Nutritional and Sensory Characteristics of Gluten-Free Quinoa (*Chenopodium Quinoa* Willd)-Based Cookies Development Using an Experimental Mixture Design. *J. Food Sci. Technol.* **2015**, *52*, 5866–5873. [CrossRef] [PubMed]
3. Coelho, M.; Salas-Mellado, M. Chemical characterization of chia (*Salvia hispanica* L.) for cse in food products. *J. Food Nutr. Res.* **2014**, *2*, 263–269. [CrossRef]
4. Miranda-Ramos, K.; Millán-Linares, M.C.; Haros, C.M. Effect of Chia as Breadmaking Ingredient on Nutritional Quality, Mineral Availability, and Glycemic Index of Bread. *Foods* **2020**, *9*, 663. [CrossRef] [PubMed]
5. AACC. *Baking Quality of Cookie Flour–Micro Method 10-52*, 9th ed.; The American Association of Cereal Chemists: St. Paul, MN, USA, 1995.
6. AOAC. Method 925.09, 991.43, 996.11, 986.11. In *Official Methods of Analysis*, 15th ed.; Association of Official Analytical Chemists: Arlington, VA, USA, 1996.
7. ISO/TS. *Food Products—Determination of the Total Nitrogen Content by Combustion According to the Dumas Principle and Calculation of the Crude Protein Content—Part 1 and 2: Cereals, Pulses and Milled Cereal Products, ISO/TS 16634-1 and ISO/TS 16634-2*; International Organization for Standardization (ISO): Geneva, Switzerland, 2016.
8. AACC. *Approved Methods of AACC: Method 08-03, 30-10*, 9th ed.; The American Association of Cereal Chemists: St. Paul, MN, USA, 2000.
9. Iglesias-Puig, E.; Monedero, V.; Haros, M. Bread with Whole Quinoa Flour and Bifidobacterial Phytases Increases Dietary Mineral Intake and Bioavailability. *LWT-Food Sci. Technol.* **2015**, *60*, 71–77. [CrossRef]
10. International Agency for Research on Cancer. *IARC Monographs on the Identification of Carcinogenic Hazards to Humans*; World Health Organization: Geneva, Switzerland, 2019.
11. Commission Regulation (EU) 2017/2158. Establishing mitigation measures and benchmark levels for the reduction of the presence as acrylamide in food. *J. Eur. Union* **2017**, *60*, 24–44.

biology and life sciences forum

Proceeding Paper

Characterization and Agronomic Evaluation of Chia Germplasm in La Plata, Buenos Aires, Argentina †

María Emilia Rodríguez [1], Ramiro Ignacio Lobo-Zavalía [2], Martín Moisés Acreche [3], Vanesa Castaldo [4], Maximiliano Pérez [4], Aline Schneider-Teixeira [1], Lorena Deladino [1] and Vanesa Yanet Ixtaina [1,5,*]

1 Centro de Investigación y Desarrollo en Criotecnología de los Alimentos (CIDCA), Consejo Nacional de Investigaciones Científicas y Técnicas (CONICET) La Plata, Universidad Nacional de La Plata (UNLP), 47 and 116, La Plata 1900, Buenos Aires, Argentina
2 Estación Experimental Agroindustrial Obispo Colombres (EEAOC), Ministerio de Desarrollo Productivo del Gobierno de Tucumán, William Cross 3150, Las Talitas 4101, Tucumán, Argentina
3 Estación Experimental Agropecuaria Salta, Instituto Nacional de Tecnología Agropecuaria (INTA), Consejo Nacional de Investigaciones Científicas y Técnicas (CONICET), Ruta Nacional 68, km 172, Cerrillos 4403, Salta, Argentina
4 Estación Experimental Gorina, Ministerio de Desarrollo Agrario de la Provincia de Buenos Aires, 501 and 149, Gorina, La Plata 1900, Buenos Aires, Argentina
5 Facultad de Ciencias Agrarias y Forestales, Universidad Nacional de La Plata (UNLP), 60 and 119, La Plata 1900, Buenos Aires, Argentina
* Correspondence: vanesaix@hotmail.com
† Presented at the IV Conference Ia ValSe-Food CYTED and VII Symposium Chia-Link, La Plata and Jujuy, Argentina, 14–18 November 2022.

Abstract: Chia (*Salvia hispanica* L.), an ancestral crop currently revalued for its nutritional properties, is one of the main sources of omega-3 fatty acids. It is a short-day plant and sensitive to frost. The objective of this study was to assess the possibility of growing chia in La Plata (Buenos Aires, Argentina) (34°54′29″ S, 58°2′25″ W) by analyzing interpopulation variability between four accessions of this species. Chia was sown on February 11 in a randomized complete block design (three replicates). Ten uniform and representative plants per plot were labeled and monitored throughout the crop cycle. Phenological stages and morphological characteristics (plant height, width of the main stem, number of pairs of folded leaves and number of pairs of side shoots, and length of central inflorescence) were recorded from seedling emergence to harvest. The emergence and the first pair of unfolded leaves were recorded 3 and 10 days after sowing, respectively. The beginning of verticillaster emergence was detected after 50 days and the beginning of flowering after 66 days of sowing. The highest growth rates were achieved in CMP and EMP. After 77 days of sowing, CMP presented the highest values for the width of the main stem (10.4 mm) and height (90.89 cm), which were statistically different from EMP. The lowest variability between populations was found for the number of pairs of unfolded leaves and side shoots. The observed variability is promising for plant breeding to obtain cultivars capable of completing their cycle at this latitude.

Keywords: crop; phenology; *Salvia hispanica* L.; variability

Citation: Rodríguez, M.E.; Lobo-Zavalía, R.I.; Acreche, M.M.; Castaldo, V.; Pérez, M.; Schneider-Teixeira, A.; Deladino, L.; Ixtaina, V.Y. Characterization and Agronomic Evaluation of Chia Germplasm in La Plata, Buenos Aires, Argentina. *Biol. Life Sci. Forum* **2022**, *17*, 16. https://doi.org/10.3390/blsf2022017016

Academic Editors: Norma Sammán, Mabel Cristina Tomás, Loreto Muñoz and Claudia Monika Haros

Published: 28 October 2022

1. Introduction

Salvia hispanica L., known as chia, is an annual herbaceous plant belonging to the *Lamiaceae* family. This species is native to southern Mexico and northern Guatemala. Currently, its cultivation is carried out commercially in Argentina, Australia, Bolivia, Colombia, Ecuador, Guatemala, Mexico, Nicaragua, Paraguay, Peru, and Southeast Asia [1,2]. Chia is originally a short-day plant, beginning the reproductive phase when the shortening of the day exceeds a certain threshold (10.5–12 h) [1,3]. It is a frost-sensitive crop. It develops optimally in tropical and subtropical climates, growing appropriately at latitudes ranging from approximately 20° N to 25° S. At higher latitudes, their growth cycle is not completed as the plants die due to frost damage [4]. Cultivars with different responses to photoperiod, capable of flowering in places where the length of the day is longer than 12.5 h, have been developed. This fact has allowed the development of its cultivation in central Europe and southern Germany [5]. In general, the grain yield is around 1400–2200 kg/ha. In addition, the content of lipids, protein, fiber, and ashes of chia fruits (nutlets, commonly named "seeds") varies between 25–32%, 19–29%, 27–29%, and 4–5%, respectively, in addition to the presence of tocopherols and polyphenols (myricetin, chlorogenic and caffeic acids, kaempferol, and quercetin) [6]. These attributes make it a crop with outstanding potential for its introduction into the horticultural systems of Gran La Plata (Buenos Aires, Argentina, 34°54′29″ S, 58°2′25″ W). This productive area has a total area of 8600 ha for horticultural production—including field and greenhouse crops—and is one of the main horticultural belts in Argentina. Artichoke (*Cynara scolymus* L.), tomato (*Solanum lycopersicum* L.), pepper (*Capsicum annuum* L.), celery (*Apium graveolens*), and lettuce (*Lactuca sativa* L.) are the main crops grown in this region [7]. Thus, the objectives of this work are to evaluate the potential of incorporating chia crop into the horticultural production systems of Gran La Plata and to analyze the interpopulation variability of four accessions of chia from our germplasm work collection.

2. Materials and Methods

2.1. Vegetal Material

The vegetal material consisted of four genotypes (populations): a commercial population cultivated by growers of Salta with mixed seeds (white, beige, and grayish brown) (CMP); one population bred by mass selection in the Estación Experimental Agroindustrial Obispo Colombres (Tucumán) (EWP1); and two populations provided by EEA INTA Salta, with white seeds (EWP2) and mixed seeds (EMP).

2.2. Edaphoclimatic Conditions

The assay was performed in the Estación Experimental MDA Gorina, located in La Plata, Buenos Aires, Argentina (34°54′29″ S, 58°2′25″ W), during the 2022 season (austral summer). In this area, the weather is of the mild-wet type, with annual rainfall of 1079 mm/year, relative humidity of around 77%, and a medium temperature of 15 °C. According to hydric balance, the weather in the region is B2C"2"r"a"(Thornthwaite system), corresponding to humid weather (B2), and microthermal (C2), indicating evapotranspiration values higher than 570 mm, with null or little water deficiency ("r") [8]. Figure 1a,b shows the mean, minimum, and maximum temperatures and the solar radiation in La Plata, respectively, during the crop cycle of chia (February to June) of 2022. These data were recorded by Weather Station OMXH®.

The soils of Gorina's series are taxonomically typical Hapludert. They are deep soils, moderately well drained, with slow or very slow permeability. They have a dark horizon of 20–30 cm in thickness, with a silty clay loam texture and clay tenors of 32–40%. The Bt horizon has a thickness between 80 and 100 cm with a clay content of 50–60%. They are fertile soils from the chemical point of view, generally provided by organic matter and nitrogen, with a slightly acidic reaction on their surface and are lightly alkaline in depth, with no interchangeable sodium harmful tenors or soluble salts [9].

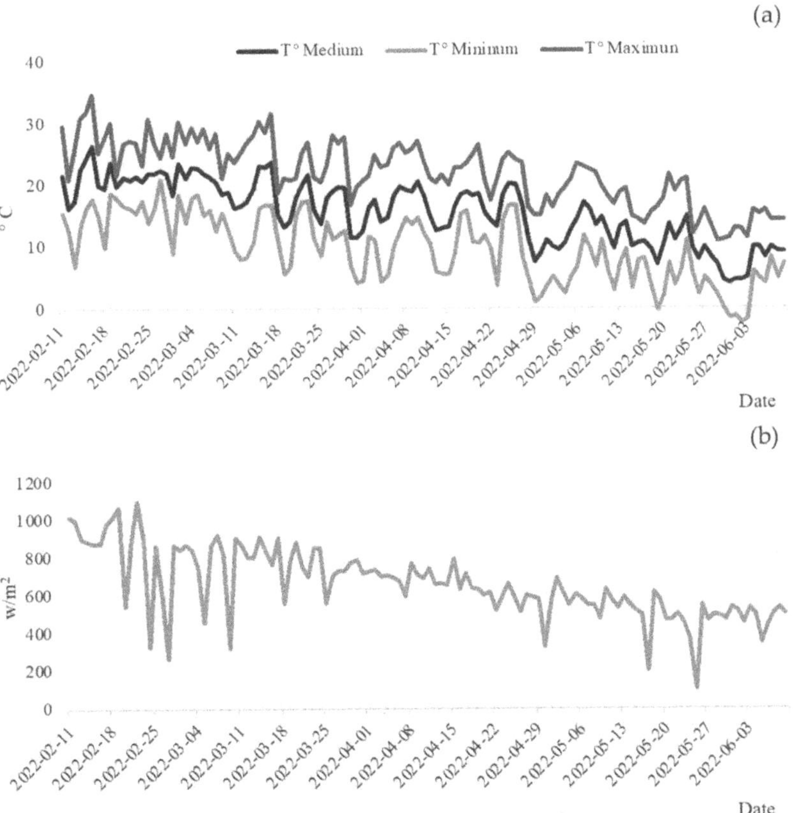

Figure 1. (**a**) Temperature and (**b**) solar radiation in La Plata (Buenos Aires, Argentina) during the chia crop cycle.

2.3. Field Experiments

The experiment was conducted on a natural field in a randomized completed block design with three replicates. The sowing date was on February 11, 2022. After 15 and 30 days of emergence, plots (four rows, 0.70 m apart, and 6 m in length) were over-seeded and hand-thinned to a final density of 8 plants per m^{-2}. To complement natural rainfall and avoid water deficits, plots were drip-irrigated for the whole of the growing seasons. The control of weeds, insects, and fungi was enhanced under organic management practices. Spontaneous species were hand removed throughout the growing season. A garlic extract solution was applied as a preventive treatment. After emergence, ten uniform plants from the central lines of each plot were labeled and monitored twice a week throughout the crop cycle.

The quantitative characteristics were measured at 33, 45, 60, 69, 76, and 83 days after sowing, recording: plant height, main stem width (at the height of the first leaf node), number of unfolded leaves on the main stem, number of pairs of side shoots, and central inflorescence length. In each plot, the days from sowing to each of the following phenological stages were also recorded: emergency, first pair of unfolded leaves, fourth pair of unfolded leaves, side shoot appearance, verticillaster appearance, flowering initiation, and full flowering. The phenological scale corresponded to the Biologische Bundesanstalt, Bundessortenamt and Chemical Industry (BBCH) systems.

2.4. Statistical Analysis

Experiments were performed following a randomized block experimental design as previously described. Data were subjected to analysis of variance (ANOVA) at a 95% confidence level ($p \leq 0.05$). Significantly different datasets were classified after post-hoc comparison tests using Tukey´s honestly significant difference test (HSD = $p \leq 0.05$). The statistical analyses were performed using the InfoStat software version 2020I (Universidad Nacional de Córdoba, Córdoba, Argentina) [10].

3. Results and Discussion

Table 1 shows the evolution of the plant height and the main stem width along the crop cycle. The CMP genotype evidenced the highest growth, recording a height of 92.15 cm after 83 days from the sowing, which was significantly different ($p \leq 0.05$) from EMP and EWP2. Additionally, CMP presented a stem width significantly higher than EMP, recording a value of 10.79 mm at the same time. The genotype EWP1 showed an intermediate behavior between EMP and CMP, with a height of 90.69 cm and a stem width of 10.44 mm after 83 days of sowing. The EMP and EWP2 populations presented a final height of 85.57 and 85.35 cm, respectively, without significant differences ($p > 0.05$) between them. Regarding the main stem width, the recorded values were 9.48 and 10.53 mm for EMP and EWP2, respectively (Table 1).

Table 1. Height of the plant (cm) and main stem width (mm) of four populations of chia (*Salvia hispanica* L.) evaluated along the crop cycle.

Character	Days after Sowing	EMP	CMP	EWP1	EWP2
Plant height (cm)	33	15.58 ^{A} ± 2.52	17.43 ^{AB} ± 2.48	18.07 ^{B} ± 3.29	18.38 ^{B} ± 3.38
	45	25.62 ^{A} ± 4.38	28.43 ^{AB} ± 4.01	28.98 ^{B} ± 5.02	29.00 ^{B} ± 5.64
	60	49.37 ^{A} ± 8.16	55.15 ^{B} ± 6.56	55.72 ^{B} ± 8.03	52.43 ^{AB} ± 9.49
	69	68.92 ^{A} ± 8.94	76.16 ^{B} ± 7.82	74.24 ^{AB} ± 9.40	70.97 ^{AB} ± 10.61
	76	83.37 ^{A} ± 8.99	90.89 ^{B} ± 9.01	87.85 ^{AB} ± 7.79	83.67 ^{A} ± 8.95
	83	85.57 ^{A} ± 9.50	92.15 ^{B} ± 9.09	90.69 ^{AB} ± 3.38	85.35 ^{A} ± 10.10
Main stem width (mm)	33	3.33 ^{A} ± 0.78	3.76 ^{AB} ± 0.76	3.45 ^{A} ± 0.75	4.15 ^{B} ± 1.49
	45	5.61 ^{A} ± 0.92	6.34 ^{B} ± 1.19	5.85 ^{AB} ± 0.91	6.28 ^{AB} ± 1.27
	60	7.55 ^{A} ± 1.34	8.41 ^{B} ± 0.60	7.74 ^{AB} ± 0.94	8.18 ^{AB} ± 1.52
	69	8.38 ^{A} ± 1.80	9.06 ^{A} ± 1.16	8.77 ^{A} ± 1.13	9.06 ^{A} ± 1.57
	76	9.25 ^{A} ± 1.46	10.40 ^{B} ± 1.41	9.85 ^{AB} ± 0.74	9.99 ^{AB} ± 1.83
	83	9.48 ^{A} ± 1.63	10.79 ^{B} ± 1.27	10.44 ^{AB} ± 1.04	10.53 ^{AB} ± 2.23

Different letters in each row indicate significant statistical differences ($p \leq 0.05$) between genotypes.

Tables 2 and 3 show the number of unfolded leaves on the main stem and the number of pairs of side shoots. These characteristics were the most stable, with no significant differences ($p > 0.05$) between the four populations studied. After 69 days of sowing, the final numbers of unfolded leaves and side shoots were 10 and 9, respectively. It is possible to observe from Tables 2 and 3 that the highest increase in these two characteristics was recorded between 33 and 45 days after sowing.

Table 2. Number of unfolded leaves on the main stem of four populations of chia (*Salvia hispanica* L.) evaluated along the crop cycle.

Days after Sowing	EMP	CMP	EWP1	EWP2
33	4.43 ^{AB} ± 0.50	4.73 ^{B} ± 0.52	4.27 ^{A} ± 0.45	4.73 ^{B} ± 0.78
45	7.02 ^{B} ± 0.65	7.27 ^{B} ± 0.55	6.50 ^{A} ± 0.51	7.23 ^{B} ± 0.54
60	9.07 ^{A} ± 0.65	9.75 ^{BC} ± 0.58	9.18 ^{AB} ± 0.59	9.76 ^{C} ± 0.73
69	10.20 ^{A} ± 0.77	10.33 ^{A} ± 0.62	9.87 ^{A} ± 0.64	10.33 ^{A} ± 0.72

Different letters in each row indicate significant statistical differences ($p \leq 0.05$) between genotypes.

Table 3. Number of pairs of side shoots of four populations of chia (*Salvia hispanica* L.) evaluated along the crop cycle.

Days after Sowing	EMP	CMP	EWP1	EWP2
33	1.93 ^{AB} ± 1.11	2.53 ^{B} ± 0.86	1.60 ^{A} ± 1.16	2.57 ^{B} ± 1.22
45	5.25 ^{A} ± 0.79	5.45 ^{A} ± 0.60	5.12 ^{A} ± 0.62	5.49 ^{A} ± 0.69
60	7.45 ^{AB} ± 0.51	7.97 ^{BC} ± 0.62	7.33 ^{A} ± 0.49	8.29 ^{C} ± 0.84
69	8.73 ^{AB} ± 0.70	9.27 ^{B} ± 0.70	8.53 ^{A} ± 0.64	8.80 ^{AB} ± 0.56
76	8.83 ^{A} ± 0.71	9.27 ^{A} ± 0.70	8.87 ^{A} ± 0.74	9.10 ^{A} ± 0.83

Different letters in each row indicate significant statistical differences ($p \leq 0.05$) between genotypes.

Finally, regarding the reproductive stage, the length of the main inflorescence had no significant differences ($p > 0.05$) between the populations studied, increasing from 18.04–20.42 mm to 138.11–156.63 mm, from 69 to 97 days after sowing (Table 4).

Table 4. Central inflorescence length (mm) from four populations of chia (*Salvia hispanica* L.) evaluated along the crop cycle.

Days after Sowing	EMP	CMP	EWP1	EWP2
69	18.04 ^{A} ± 4.66	20.42 ^{A} ± 3.74	18.20 ^{A} ± 4.06	20.08 ^{A} ± 6.10
74	35.41 ^{A} ± 8.09	36.69 ^{A} ± 9.83	33.49 ^{A} ± 8.01	36.69 ^{A} ± 13.40
76	45.21 ^{A} ± 11.62	46.86 ^{A} ± 10.47	39.71 ^{A} ± 10.78	45.70 ^{A} ± 17.16
83	59.30 ^{A} ± 18.19	64.27 ^{A} ± 16.37	55.53 ^{A} ± 18.55	63.81 ^{A} ± 27.99
90	116.31 ^{A} ±35.87	118.08 ^{A} ± 26.01	108.70 ^{A} ± 38.89	110.35 ^{A} ± 38.91
97	156.63 ^{A} ±32.76	138.11 ^{A} ± 36.71	155.08 ^{A} ± 34.58	141.05 ^{A} ± 32.74

Different letters in each row indicate significant statistical differences ($p \leq 0.05$) between genotypes.

Regarding the principal growth stages, six of the ten stages described in the BBCH scale were achieved by the chia crop in the Gran La Plata region under field conditions: germination (stage 0), leaf appearance (stage 1), shoot appearance (stage 2), stem elongation (stage 3), inflorescence growth on the main stem verticillaster (stage 5), and flowering (stage 6).

Figure 2 shows the evolution of the chia crop, indicating the phenological stage according to the scale proposed by the Biologische Bundesanstalt, Bundessortenamt and Chemical Industry (BBCH) and applied by Brandán et al. [11].

Figure 2. Crop evolution along the vegetative and reproductive cycles of chia. Numbers indicate phenological stages according to the Biologische Bundesanstalt, Bundessortenamt and Chemical Industry (BBCH) system. (**a**) growth stage 12, second pair (four leaves) unfolded; (**b**) growth stage 14, fourth pair (eight leaves) unfolded; (**c**) growth stage 22, four side shoots detectable; (**d**) growth stage 28, eight side shoots detectable; (**e**) growth stage 65, full flowering: at least one open flower in the apical-third of the verticillaster; and (**f**) growth stage 71, grains of the basal-third of the verticillaster with milk texture.

Overall, the emergence, recorded when cotyledons emerge over the soil surface (stage 09 of the BBCH scale), was verified 3 days after sowing. This stage is relevant in crop establishment as any environmental adversity could generate a low plant stand [11]. The first pair of leaves (stage 11) was recorded after 10 days from sowing. Figure 2a,b show stages 12 and 14, respectively, which correspond to the second (four leaves) and four pairs (eight leaves) of unfolded leaves. In this trial, the final number of unfolded leaves was about 10–11 for all the genotypes. Stage 14 was recorded 31 days after sowing, which coincided with the appearing of the first pair of side shoots (stage 21), which present an acropetal growth.

The number of days to differentiate verticillaster from leaves (stage 51) was 53 days from sowing when plants presented 10–11 unfolded pairs of leaves (stages 110 and 11), and 8–9 pairs of side shoots (stages 28 and 29).

The beginning of flowering (stage 60) was 66 days after sowing, when at least one flower in the basal-third of the verticillaster had opened. Full flowering (stage 65) was achieved in CMP and EMP at 76 days after sowing, whereas EWP1 and EWP2 took 86 days, indicating that the first two genotypes greater were earlier than the last two.

4. Conclusions

The main differences between the assayed populations were detected for the height and the stem diameter; the number of leaves and shoots were stable regardless of the population evaluated. The plants that first reached the flowering stage corresponded to genotypes with mixed seeds, thus being the earliest ones. The first frost recorded in the area was on April 29 (Figure 1a). After that date, the medium temperatures dropped along with solar radiation (Figure 1b), hindering the favorable evolution because it is a frost-sensitive crop. The assay was finished 100 days after sowing with plants highly affected by low temperatures, and genotypes could not complete the cycle under field conditions. However, it is relevant to note that the populations were also evaluated under greenhouse conditions (data not shown). Under these conditions, they did not show cold damage and were in grain filling after 160 days from the sowing date.

The introduction of genotypes less sensitive to the photoperiod is necessary for the adoption of this crop in the Gran La Plata area if it is to be produced in the field. However, future research on the results obtained under greenhouse conditions may be potentially favorable for incorporating the crop into covered horticultural systems.

Variability was found among the studied populations, which is promising for future genetic breeding plans.

Author Contributions: Conceptualization, M.E.R. and V.Y.I.; methodology, M.E.R.; software, L.D.; formal analysis, M.E.R. and V.C.; investigation; resources, A.S.-T., V.Y.I., R.I.L.-Z. and M.M.A.; data curation, M.E.R.; writing—original draft preparation, M.E.R. and L.D.; writing—review and editing, V.Y.I.; supervision, M.P., V.Y.I.; project administration, V.Y.I. and A.S.-T.; funding acquisition, A.S.-T. and V.Y.I. All authors have read and agreed to the published version of the manuscript.

Funding: This work was supported by the following grants: Ia ValSe-Food-CYTED (119RT0567), PICT 2018-2352, PICT 2020-191; and UNLP X-901.

Informed Consent Statement: Not applicable.

Acknowledgments: The authors express thanks to the technical personnel of Estación Experimental del MDA Gorina, the Estación Experimental Agroindustrial Obispo Colombres (EEAOC), EEA Salta, and INTA-CONICET who provided chia seeds.

Conflicts of Interest: The authors declare no conflict of interest.

References

1. Ayerza, R.; Coates, W. *Chia: Rediscovering a Forgotten Crop of the Aztecs*; University of Arizona Press: Tucson, AZ, USA, 2005.
2. Orona-Tamayo, D.; Valverde, M.; Paredes-Lopez, O. Chia—The new golden seed for the 21st century: Nutraceutical properties and technological uses. In *Sustainable Protein Sources*; Elsevier: Amsterdam, The Netherlands, 2017; pp. 265–281.
3. Hildebrand, D.; Jamboonsri, W.; Phillips, T. Early Flowering Mutant Chia and Uses Thereof. U.S. Patent No. 8,586,831, 19 November 2013.
4. Sorondo, A. Chia (*Salvia hispanica* L.) variety Sahi Alba 914. Unites States Patent Aplication Publication. U.S. Patent No. 2014/0325694-A1, 27 July 2014.
5. Grimes, S.J. Screening and Cultivation of Chia (*Salvia hispanica* L.) under Central European Conditions: The potential of a Re-Emerged Multipurpose Crop. 2021. Available online: http://opus.uni-hohenheim.de/volltexte/2021/1972/pdf/Druckreife_SamanthaJoGrimes_Print_Thesis.pdf (accessed on 13 October 2022).
6. Ixtaina, V.Y.; Martínez, M.L.; Spotorno, V.; Mateo, C.M.; Maestri, D.M.; Diehl, B.W.; Nolasco, S.M.; Tomás, M.C. Characterization of chia seed oils obtained by pressing and solvent extraction. *J. Food Comp. Anal.* **2011**, *24*, 166–174. [CrossRef]
7. García, M.; Quaranta, G.J.G. Análisis de las estadísticas hortícolas de Buenos Aires. *Un aporte para la cuantificación de los establecimientos hortícolas de La Plata. Geograficando* **2022**, *18*, e108.
8. Kruse, E.; Sarandón, R.; Gaspari, F. *Impacto del cambio climático en el Gran La Plata*; Edulp, Editorial de la Universidad de La Plata: La Plata, Argentina, 2014.
9. Hurtado, M.A.; Giménez, J.E.; Cabral, M.G.; Silva, M.M.; Martinez, O.R.; Camilión, M.C.; Sanchez, C.A.; Muntz, D.; Gebhard, J.A.; Forte, L.M.; et al. Análisis Ambiental del Partido de La Plata. 2006. Available online: http://sedici.unlp.edu.ar/handle/10915/27046 (accessed on 14 July 2022).

10. Di Rienzo, J.A.; Casanoves, F.; Balzarini, M.G.; González, L.; Tablada, M.; Robledo, C.W. InfoStat: Programa de Cómputo, versión 24-03-2011. Available online: http://www.infostat.com.ar/ (accessed on 1 July 2022).
11. Brandán, J.P.; Curti, R.N.; Acreche, M.M. Phenological growth stages in chia (*Salvia hispanica* L.) according to the BBCH scale. *J. Sci. Hortic.* **2019**, *255*, 292–297. [CrossRef]

Proceeding Paper

Effect of Cooking on the Content and Bioaccessibility of Minerals in Pseudocereals †

Carla Motta [1,*], Isabel Castanheira [1], Ana Sofia Matos [2], Ana Claudia Nascimento [1], Ricardo Assunção [3,4], Carla Martins [5,6] and Paula Alvito [1,5]

1 Departamento de Alimentação e Nutrição, Instituto Nacional de Saúde Doutor Ricardo Jorge, INSA, IP, Avenida Padre Cruz, 1649-016 Lisboa, Portugal
2 Departamento de Engenharia Mecânica e Industrial, UNIDEMI, Faculdade de Ciências e Tecnologia, Universidade Nova de Lisboa, 2829-516 Caparica, Portugal
3 Instituto Universitário Egas Moniz—Cooperativa de Ensino Superior Campus Universitário, Quinta da Granja, Monte de Caparica, 2829-511 Caparica, Portugal
4 CESAM-Centre for Environmental and Marine Studies, University of Aveiro, Campus Universitário de Santiago, 3810-193 Aveiro, Portugal
5 Public Health Research Centre, NOVA National School of Public Health, Universidade NOVA de Lisboa, Avenida Padre Cruz, 1600-560 Lisbon, Portugal
6 Comprehensive Health Research Center, Campo Mártires da Pátria, Universidade NOVA de Lisboa, 1169-056 Lisbon, Portugal
* Correspondence: carla.motta@insa.min-saude.pt
† Presented at the IV Conference Ia ValSe-Food CYTED and VII Symposium Chia-Link, La Plata and Jujuy, Argentina, 14–18 November 2022.

Abstract: The cooked Andean cereals can be considered a good source of minerals, contributing to the recommended daily intakes as observed in previous works. This study evaluated quinoa, amaranth, and buckwheat's chemical and nutritional compositions and their bioaccessibility through an in vitro gastric digestion simulation to understand their dietary changes. ICP-OES was used to quantify the mineral profile, and the impact of cooking on bioaccessibility was evaluated using multivariate statistical analysis. In this context, the contents of some essential minerals (potassium, magnesium, calcium, zinc, copper, iron and manganese) were evaluated. The lowest cooking losses were noted for calcium in quinoa (67%), and the highest was found for zinc in buckwheat (73%). The calcium and manganese concentration varied considerably with boiling among Andean cereals. For copper, magnesium, iron and manganese, was observed a higher bioaccessibility in cooked quinoa and amaranth. The lowest bioaccessibility was detected for phosphorus in the boiled quinoa fraction (36%). The results highlight the need to consider the losses in bioavailability for minerals during digestion and the related influence on the estimation of proper nutrient intake. These results contribute to understanding the bioaccessibility of minerals in cooked Andean cereals and the changes in these nutrient contents through the boiling process. Other ongoing cooking processes lead to a scientific recommendation of the best cooking method for boosting nutrient intake.

Keywords: amaranth; bioaccessibility; buckwheat; quinoa; minerals; nutrient intake

Citation: Motta, C.; Castanheira, I.; Matos, A.S.; Nascimento, A.C.; Assunção, R.; Martins, C.; Alvito, P. Effect of Cooking on the Content and Bioaccessibility of Minerals in Pseudocereals. *Biol. Life Sci. Forum* **2022**, *17*, 17. https://doi.org/10.3390/blsf2022017017

Academic Editors: Norma Sammán, Mabel Cristina Tomás, Loreto Muñoz and Claudia Monika Haros

Published: 23 October 2022

1. Introduction

Dietary Guidelines recommend Andean cereal consumption worldwide, recognising their favourable nutrient profile. Moreover, international organisations such as the Food and Agriculture Organization recommend Andean Cereals as staple foods to fulfil the human diet's essential protein and energy requirements [1]. The growing interest of Western Diets in Andean cereals, especially Quinoa Amaranth and Buckwheat, as ingredients to improve the nutritional value of plant-based pathways and to create novel foodstuffs to replace meat. The value of Quinoa and Amaranth as nutritious food is due to the highest combined amino acid profile with dietary fibre and minerals [2,3] quinoa and amaranth have a significant amount of anti-nutrients like saponins and phytates, which can have contradictory effects depending on the amount. Saponins have been shown to interfere with the absorption of micronutrients by displaying anti-enzymatic activities. Phytates can have a chelating impact at higher doses, decreasing minerals' bioavailability [4]. Published Data on Andean Cereal does not provide sufficient information to highlight the relevance of Andean Cereals mainly due to the unknown effect of anti-nutrients on cooked foods and their impact on the human body after digestion.

The standardisation of analytical and in vitro digestion procedures is a strategy accepted by the scientific community to define the influence of food processing on food composition data. Household processing such as boiling reduces the anti-nutrients effects of saponins and phytates, improving the mineral bioavailability [5]. Therefore, accurate determination of the component values (of foods as consumed) is achieved using a combination of analytical determination with in vitro studies [6,7].

The present study addresses the question by comprehensively describing the mineral profile of the buckwheat, quinoa and amaranth, comparing the bioavailability of raw material with boiled under traceable measurements procedures to guarantee a deep inside estimation of mineral nutrient intakes.

2. Materials and Methods

2.1. Chemicals

All reagents are of standard analytical grade. For mineral analysis, hydrogen peroxide (30%) and supra pure nitric acid (65%), ultrapure grade, were purchased from Merck (Darmstadt, Germany). Standards of each mineral, with a concentration of 1000 mg/L, in trace element grade nitric acid 2–3%, were purchased from SCP SCIENCE (Courtaboeuf, France). All calibration curves, diluted in nitric acid at 2%, ranged from 0.02–0.2 mg/L for Cu and Mn, 0.05–0.5 for Fe and Zn, 1–10 mg/L for Mg, 2–20 mg/L for Ca, P, Na and 2.5–25 mg/L for K. For bioaccessibility assessment all reagents are standard analytical grade. The different enzymes (Human salivary α-amylase; Pancreatin; Porcine pepsin, and Bile) and Pefabloc SC (4-(2-Aminoetyl) benenesulfonyl fluoride) were provided by Sigma Aldrich (St Louis, MO, USA). Salts such as calcium chloride anhydrous; sodium hydroxide; sodium chloride; magnesium chloride; sodium bicarbonate; ammonium carbonate; potassium phosphate monobasic; potassium chloride, and hydrochloric acid, used to perform the different digestion phases, where provided by Merck (Darmstadt, Germany).

2.2. Sampling and Sample Preparation

White amaranth (*Amaranthus* sp.), buckwheat (Fagopyrum esculentum Moench) and quinoa (*Chenopodium* sp.) are from biological agriculture collected in Lisbon markets. The sampling plan, including cooking methods and samples, was defined according to the protocol described in Motta et al. (2019) [2]. All samples were washed before any procedure. Boiling was performed in 50 g of raw seeds with 150 g of ultrapure water in a Termomix® TM31 food processor (Vorwerk, Wuppertal, Germany), set to 100 °C during 15 min. All raw and cooked grains were ground in a GRINDOMIX GM 200 high-speed grinder (Retsch, Düsseldorf, Germany) and stored in vacuum bags at 4 °C until use.

2.3. Moisture and Mineral Analysis

Moisture Content Was Determined According to AOAC (AOAC 952.08, 2000) in a Dry Air oven (102 °C ± 2 °C) until Constant Weight as described in our previous work [2]. Briefly, each sample was tested in quadruplicate for mineral quantification, using 0.5 g of sample to digest in a closed vessel microwave digestion system (Milestone ETHOS 1 Series, Shelton, CT, USA) under acid hydrolysis with nitric acid and hydrogen peroxide. Minerals were analysed by an inductively coupled plasma optical emission spectrometer, ICP-OES iCAP 6000 series (Thermo Fisher Scientific, Waltham, MA, USA), with radial and axial configuration and quantified by an external calibration curve. As part of the quality control, two independent samples were analysed in each group of 10 samples with an acceptance criterion of ±10%.

2.4. In Vitro Gastrointestinal Digestion

To access minerals bioaccessibility, a portion of one gram of each sample was weight. All experiments were conducted in triplicate, and a reagent blank was performed in every batch of samples. Bioaccessibility studies were performed according to the harmonised static in vitro digestion (IVD) model described by [6,7]. This model includes three sequential phases: oral phase (simulated saliva fluid with amylase—75 U/mL, pH 7), gastric phase (simulated gastric fluid with pepsin—2000 U/mL, pH 3) and intestinal phase (simulated intestinal fluid containing a pancreatin-bile mixture—100 U/mL and 10 mM, respectively, pH 7). After intestinal phase incubation, the reaction was stopped by adding 1 mM of Pefabloc® (Sigma-Aldrich, St. Louis, MO, USA). Digests of sample seeds and blanks were immediately placed in liquid nitrogen, and after that, samples were kept at −80° C. The mineral content was quantified in the blank, and seed samples digest fluids. Blank quantification was deducted from the seed sample digested extract quantification.

Bioaccessibility values were determined according to Equation (1):

$$\text{Bioaccessibility (\%)} = (\text{[mineral] in digestion extract})/(\text{[mineral] in non-digested sample}) \quad (1)$$

2.5. Statistical Analysis

Results were expressed as mean and standard deviation (SD). The comparison of means was analysed by analysis of variance (ANOVA) and Tukey's test ($p < 0.05$) using Statistica v. 8 software (Statsoft Ibérica, Lisboa, Portugal).

3. Results

3.1. Moisture and Mineral Content

Mineral content for raw and cooked pseudocereals is present in Table 1. The water content was increased from 55% to 63% in boiled amaranth buckwheat and quinoa, where amaranth was the seed with the higher water retention. Raw amaranth also presents the highest content in all minerals except copper and potassium. In all cases, as expected, the mineral content was reduced after the seeds were cooked. Considering pseudocereals as consumed, the highest concentration was obtained for potassium in boiled quinoa (182 mg/100 g) and for phosphorus in boiled amaranth (175 mg/100 g), while the lowest mineral activity was found for copper in boiled buckwheat.

Table 1. Mineral and moisture content in raw and boiled pseudocereals mg/100 g (fresh weight) [1].

Food	Process	Cu	Mn	Fe	Zn	Mg	Ca	P	K	Moisture
Amaranth	Raw	0.485 ± 0.018	2.93 ± 0.288	6.81 ± 0.487	3.85 ± 0.184	267 ± 23.0	158 ± 19.4	505 ± 55.6	455 ± 32.5	10.4 ± 0.96
	Boiled	0.172 ± 0.004	1.20 ± 0.016	1.96 ± 0.018	1.25 ± 0.030	80.8 ± 1.01	54.5 ± 0.388	173 ± 1.56	142 ± 0.745	73.9 ± 0.83
Buckwheat	Raw	0.427 ± 0.021	1.11 ± 0.098	2.55 ± 0.161	1.98 ± 0.336	216 ± 12.9	15.3 ± 0.889	387 ± 35.6	460 ± 26.7	13.4 ± 0.15
	Boiled	0.138 ± 0.001	0.355 ± 0.011	0.830 ± 0.038	0.535 ± 0.011	66.6 ± 2.43	5.22 ± 0.148	118 ± 4.05	146 ± 2.97	68.5 ± 0.17
Quinoa	Raw	0.534 ± 0.057	2.24 ± 0.323	4.30 ± 0.461	2.97 ± 0.327	224 ± 33.0	57.9 ± 6.54	444 ± 42.1	506 ± 18.4	11.7 ± 0.17
	Boiled	0.154 ± 0.005	0.638 ± 0.052	1.49 ± 0.097	0.979 ± 0.080	64.0 ± 2.65	24.8 ± 3.39	149 ± 3.10	179 ± 2.61	66.6 ± 0.16

[1] Values as mean and standard deviation ($n = 4$).

3.2. Bioaccessibility of Minerals

The elemental bioaccessibility (%) of pseudocereals is presented in Table 2. The table shows the variation of the mineral bioaccessibility (%) in raw and boiled seeds.

Raw quinoa presented higher bioaccessibility for potassium (87–100%), while boiled quinoa's highest values appear in copper and calcium (79–100%). The bioaccessibility of the different minerals only increases after cooking, especially in amaranth. In that case, amaranth presents a higher bioaccessibility for phosphorus and potassium (78–89%). Buckwheat shows the higher bioaccessibility for zinc, either raw or boiled (61–100%), for magnesium and calcium before cooking (42–58%).

Table 2. Bioaccessibility (%) of minerals in pseudocereals after digestion (%) [1,2].

Food	Process	Cu	Mn	Fe	Zn	Mg	Ca	P	K
Amaranth	Raw	25.1 ± 4.66	13.9 ± 2.21	10.9 ± 1.75	62.4 ± 8.67	27.0 ± 6.49	50.8 ± 5.01	47.0 ± 5.67	59.2 ± 13.57
	Boiled	60.6 ± 2.20	31.3 ± 1.54	13.8 ± 0.70	45.9 ± 1.91	40.4 ± 1.64	54.4 ± 1.65	82.0 ± 3.28	83.3 ± 5.84
Buckwheat	Raw	48.9 ± 8.10	12.8 ± 1.59	21.5 ± 1.75	75.8 ± 14.08	49.6 ± 8.12	98.0 ± 8.61	54.2 ± 6.58	79.5 ± 5.43
	Boiled	53.7 ± 2.20	50.5 ± 3.59	20.5 ± 2.43	88.5 ± 5.67	63.8 ± 4.92	77.3 ± 13.5	40.0 ± 4.65	79.3 ± 0.79
Quinoa	Raw	33.9 ± 8.14	14.0 ± 2.95	23.0 ± 4.07	35.5 ± 6.59	41.1 ± 4.24	36.8 ± 7.48	52.6 ± 9.59	96.4 ± 8.71
	Boiled	85.8 ± 6.66	31.2 ± 1.08	34.7 ± 2.37	37.1 ± 1.90	63.7 ± 2.69	97.0 ± 5.64	36.0 ± 0.44	39.7 ± 2.01

[1] Values as mean and standard deviation (n = 4). [2] Values of bioaccessibility (%) were obtained by Equation (1).

Nutrient retention (NR) is related to the food matrix complexity that occurs naturally with the cooking processes. Using the bioaccessibility (%) for each nutrient in the cooked food and the correspondent bioaccessibility (%) in the raw food, we can easily, by a ratio, evaluate the impact of the boiling process on bioaccessibility. Data regarding NR impact on bioaccessibility for amaranth, buckwheat and quinoa, as shown in Figure 1. Comparing the effect of boiling on the bioaccessibility of the seeds, we can conclude that boiling can increase to 3.5 times the bioaccessibility of manganese and zinc in buckwheat, calcium and copper in quinoa and phosphorus, copper, and potassium in amaranth.

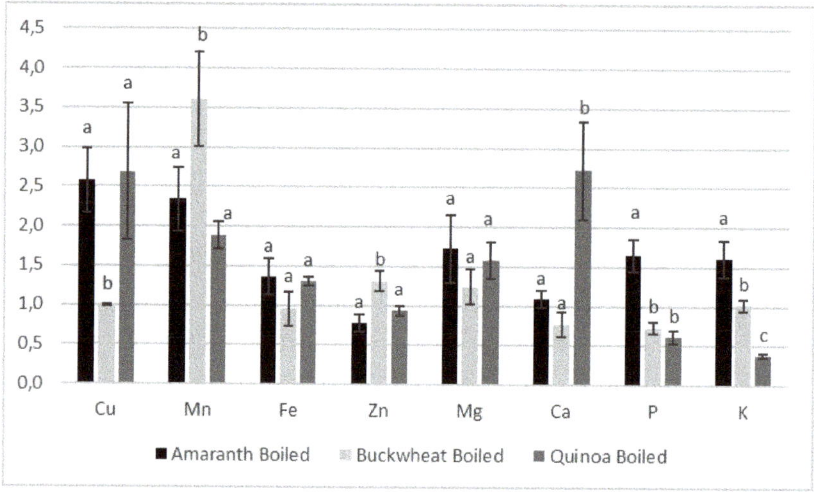

Figure 1. Effect of boiling on bioaccessibility concerning the nutrient retention (NR), using the ratio of bioacessible portion on raw and boiled seeds. Bioacessibility % raw sedd/Bioacessibility % boiled seed. Note: Different letters show statistically significantly ($p < 0.05$) differences in mineral content after boiling when compared with raw.

3.3. Contribution of Pseudocereals to the Population Intake

To evaluate the pseudocereals nutrient contribution to the Recommended Nutrient Intake (RNI) or Adequate Intakes (AI) of raw and boiled seeds, we defined, under nutritional portion/day recommendations, that one equivalent portion corresponds to two tablespoons of raw seeds (70 g), that after cooking represent 166 g of quinoa and 200 g of amaranth and buckwheat. RNI has been assessed by the World Health Organization and Food and Agriculture Organization of the United Nations WHO/FAO, for adults, males and females [8]. European Food Safety Authority [7,9] reports the AI for manganese and phosphorus for adults. Results for the bioaccessible fraction and composition are presented in Table 3.

Table 3. Contribution of one equivalent portion of pseudocereals to the population reference intakes (PRI) and adequate intakes (AI), for males and females aged over 18 years, for minerals before and after digestion.

			Cu		Mn		Fe		Zn		Mg		Ca		P		K	
PRI */AI Rference Value		M	1.6		3.0		11 *		16.3 *		350		950 *		550		3500	
		F	1.5				16 *		12.7 *		300						4000	
% Contribution			Bioac [a]	Comp [b]	Bioac [a]	Comp [b]	Bioac [a]	Comp [b]	Bioac [a]	Comp [b]	Bioac [a]	Comp [b]	Bioac [a]	Comp [b]	Bioac [a]	Comp [b]	Bioac [a]	Comp [b]
Amaranto	Raw	M	5.3	21	9.4	68	4.7	44	10	16	14	53	5.9	11	30	64	5.3	9.0
		F	5.7	23			3.2	30	13	21	17	62					4.7	7.9
	Boiled	M	13	22	25	80	4.9	36	7.0	15	19	46	6.3	11	52	63	6.8	8.1
		F	14	23			3.4	25	9.0	20	22	54					5.9	7.1
Buckwheat	Raw	M	9.1	17	3.3	26	3.5	16	6.4	8.6	21	43	1.1	1.1	27	50	7.3	9.2
		F	9.7	20			2.4	11	8.2	11	25	51					6.4	8.1
	Boiled	M	9.3	17	12	24	3.1	15	5.8	6.6	24	38	0.9	1.1	17	43	6.6	8.3
		F	9.9	18			2.1	10	7.5	8.4	28	44					5.8	7.3
Quinoa	Raw	M	7.8	23	7.1	52	6.2	27	4.5	13	18	45	1.6	4.3	29	57	9.7	10
		F	8.3	25			4.3	19	5.8	16	21	52					8.5	8.9
	Boiled	M	14	16	10.4	35	7.9	22	4.0	10	19	30	4.6	4.3	16.2	45	3.4	8.5
		F	15	17			5.4	15	5.1	13	22	35					2.9	7.4

[a]—Bioacessible fraction. [b]—Composition data. M—Male; F—Female. Intakes expressed in mg/day; * Population reference intakes.

Considering the bioaccessible portion between raw and cooked seeds, all minerals except for zinc and potassium in quinoa present a higher contribution to RNI or AI. Especially in amaranth, the boiling process promotes a significant increase in bioaccessibility, between 41% and 62% for manganese, copper, and phosphorus. In opposition, boiling seems to decrease in mean, around 23%, the zinc bioaccessibility for all seeds.

Regarding the percentage of intake supplied by one portion of cooked seeds, amaranth contributes to 52% of phosphorus and buckwheat with 26% of magnesium intake. However, pseudocereals are not good sources of calcium (3%) or iron (4%.) of the RNI.

Suppose we use the mineral composition to calculate de RNI or AI of a consumer without knowledge about the bioaccessible fraction. In that case, and using the presented results, we can induce an error in the calculations of intakes for copper, zinc, magnesium, calcium, phosphorus, and potassium, around 45% below the corrected intake, then if we consider the bioaccessible fraction. For manganese and iron, the error can be higher than 70%.

4. Discussion

The obtained results considering mineral content are aligned with previous work published by Motta et al. [10] and in agreement with retention factors for quinoa obtained in USDA [3] for pools of boiled quinoa from different sources. Low bioaccessibility of some minerals, especially in raw seeds, can occur due to phytates, known as anti-nutrients, that may strongly connect with Zn, reducing absorption by metal precipitation. Ovca et al. (2011) results in studies with pumpkin seeds report that 50–80% of phosphorus in seeds occurs as phytic acid, which may cause complex nutrients such as K, Mg, Ca, Fe, Zn and Mn reducing their bioavailability in gastrointestinal conditions. Rousseau et al. [4] also conclude that although cooking can have a positive effect, it can also reduce mineral bioaccessibility, induced by insoluble and non-absorbable mineral chelates formed with dietary fibre, phytic acid, which cannot be destroyed during the digestion process. Our work converged with concepts postulated by Rousseau et al. [4]. Knowing that cooking is relevant in enhancing bioavailability patterns in the intestine, we continue working on other processing methods to evaluate its effects on mineral bioaccessibility.

5. Conclusions

This research represents the first study for understanding the mineral bioaccessibility in boiled amaranth quinoa and buckwheat. Furthermore, we developed a standardisation in vitro method allowing the comparability of bioaccessibility fraction. This achievement is relevant for accurately estimating mineral uptake and highlighting the contribution of pseudocereals to a healthy diet.

Although our experimental data based on retention rate demonstrated that boiled pseudocereals might accumulate a considerable amount of minerals in bioaccessible fraction, further research into the mineral release mechanisms is needed to understand the effect of the anti-nutrient and membrane brake on each pseudocereal at a cellular level.

Author Contributions: Conceptualization, C.M. (Carla Motta) and I.C.; methodology, C.M. (Carla Martins); R.A. and P.A.; software, A.S.M.; validation, I.C., R.A. and P.A.; formal analysis, C.M. (Carla Motta) and A.C.N.; investigation, C.M. (Carla Motta); I.C. and P.A.; resources, C.M. (Carla Martins) and P.A.; data curation, A.S.M. and R.A.; writing—original draft preparation, C.M. (Carla Motta); writing—review and editing, I.C.; visualization, A.S.M. and C.M. (Carla Motta); supervision, I.C. and P.A.; project administration, I.C.; funding acquisition, I.C. All authors have read and agreed to the published version of the manuscript.

Funding: This work was supported by grant Ia ValSe-Food-CYTED (119RT0567).

Institutional Review Board Statement: Not applicable.

Informed Consent Statement: Not applicable.

Data Availability Statement: The data presented in this study are available on request from the corresponding author.

Acknowledgments: This work was supported by grant Ia ValSe-Food-CYTED (119RT0567).

Conflicts of Interest: The authors declare no conflict of interest.

References

1. FAO; WHO. Human Vitamin and Mineral Requirements. In *Human Vitamin and 2. Mineral Requirements*; FAO: Rome, Italy, 2001; pp. 181–194.
2. Motta, C.; Castanheira, I.; Gonzales, G.B.; Delgado, I.; Torres, D.; Santos, M.; Matos, A.S. Impact of Cooking Methods and Malting on Amino Acids Content in Amaranth, Buckwheat and Quinoa. *J. Food Compos. Anal.* **2019**, *76*, 58–65. [CrossRef]
3. US Department of Agriculture and Agricultural Service. USDA National Nutrient Database for Standard Reference, Release 28. Nutrient Data Laboratory. 2019. Available online: https://fdc.nal.usda.gov/ (accessed on 26 June 2022).
4. Rousseau, S.; CKyomugasho, l.; Celus, M.; Hendrickx, M.E.G.; Grauwet, T. Barriers Impairing Mineral Bioaccessibility and Bioavailability in Plant-Based Foods and the Perspectives for Food Processing. *Crit. Rev. Food Sci. Nutr.* **2020**, *60*, 826–843. [CrossRef] [PubMed]
5. Platel, K.; Srinivasan, K. Bioavailability of Micronutrients from Plant Foods: An Update. *Crit. Rev. Food Sci. Nutr.* **2016**, *56*, 1608–1619. [CrossRef] [PubMed]
6. Minekus, M.; Alminger, M.; Alvito, P.; Ballance, S.; Bohn, T.; Bourlieu, C.; Carrière, F.; Boutrou, R.; Corredig, M.; Dupont, D.; et al. A Standardised Static in Vitro Digestion Method Suitable for Food—An International Consensus. *Food Funct.* **2014**, *5*, 1113–1124. [CrossRef] [PubMed]
7. European Food Safety Authority. Scientific Opinion on Dietary Reference Values for Manganese. *EFSA J.* **2013**, *11*, 3419. [CrossRef]
8. Food and Agricultural Organization to the United Nations (FAO). *Sustainable Diets and Biodiversity. Biodiversity and Sustainable Diets United against Hunger*; FAO: Rome, Italy, 2010.
9. European Food Safety Authority. Scientific Opinion on Dietary Reference Values for Phosphorus. *EFSA J.* **2015**, *13*, 4185. [CrossRef]
10. Motta, C.; Nascimento, A.C.; Santos, M.; Delgado, I.; Coelho, I.; Rego, A.; Matos, A.S.; Torres, D.; Castanheira, I. The Effect of Cooking Methods on the Mineral Content of Quinoa (Chenopodium Quinoa), Amaranth (*Amaranthus* sp.) and Buckwheat (Fagopyrum Esculentum). *J. Food Compos. Anal.* **2016**, *49*, 57–64. [CrossRef]

Proceeding Paper

Red and Gray Bean (*Phaseolus vulgaris* L.) Protein Hydrolysates: Food Prototypes with Pota (*Dosidicus gigas*) by-Product Meal †

N. Chasquibol [1,*], R. Alarcón [1], B. F. Gonzales [1], M. Tapia [1], P. Jara [1], A. Sotelo [1], B. García [2] and M. C. Pérez-Camino [2]

1 Grupo de Investigación en Alimentos Funcionales, Carrera de Ingeniería Industrial, Instituto de Investigación Científica, Universidad de Lima, Av. Javier Prado Este 4600, Fundo Monterrico Chico, Surco, Lima 15023, Peru
2 Department of Characterization and Quality of Lipids, Instituto de la Grasa-CSIC, Universidad Pablo de Olavide, 41013 Seville, Spain
* Correspondence: nchasquibol@ulima.edu.pe

† Presented at the IV Conference Ia ValSe-Food CYTED and VII Symposium Chia-Link, La Plata and Jujuy, Argentina, 14–18 November 2022.

Abstract: The bean (*Phaseolus vulgaris* L.) known as the ñuña, numia, or Andean popping bean, native to the central Andes of Peru, is often consumed as a snack food after a quick toasting process. The characterization of two of its varieties, red and gray ñuña beans, was performed to determine their proximate composition, total phenolics, and antioxidant activity. Moisture content ranged from 12.67% (red ñuña bean) to 11.94% (gray ñuña bean); fat content varied from 1.77% (red ñuña bean) to 1.44% (gray ñuña bean); protein content was high, with a content range from 23.90% (red ñuña bean) to 26.81% (gray ñuña bean); ash content ranged from 4.04% (red ñuña bean) to 3.88% (gray ñuña bean); and a high content of carbohydrates was also found (from 57.60 to 55.94%). The phenolic compounds were consistently higher according to particle size, and the total phenolic content varied from 8589 μg of gallic acid equivalent (GAE)/g powder (red ñuña bean) to 3478 μg GAE/g powder (gray ñuña bean), with antioxidant activity varying from 9879 μg trolox/g powder (red ñuña bean) to 5539 μg trolox/g powder (gray ñuña bean). Food prototypes were then developed with the hydrolyzed proteins from ñuña beans, mashua *(Tropaeolum tuberosum)* tuber flour, purple corn (*Zea mays* L.) flour, and pota (*Dosidicus gigas*) by-product meal with a high content of protein and omega-3 acids (~50% eicosapentaenoic acid (EPA) + docosahexaenoic acid (DHA) on total fat).

Keywords: *Phaseolus vulgaris*; DPPH; degree of hydrolysis; protein; polyphenols; prototypes; ñuña beans

Citation: Chasquibol, N.; Alarcón, R.; Gonzales, B.F.; Tapia, M.; Jara, P.; Sotelo, A.; García, B.; Pérez-Camino, M.C. Red and Gray Bean (*Phaseolus vulgaris* L.) Protein Hydrolysates: Food Prototypes with Pota (*Dosidicus gigas*) by-Product Meal. *Biol. Life Sci. Forum* **2022**, *17*, 18. https://doi.org/10.3390/blsf2022017018

Academic Editors: Norma Sammán, Mabel Cristina Tomás, Loreto Muñoz and Claudia Monika Haros

Published: 2 November 2022

1. Introduction

Ñuña beans are an important crop and a typical food for the Andean population, especially in the regions from Cajamarca, Peru to Chuquisaca, Bolivia [1]. These beans have a high protein content of around 20%, with a low foaming and viscosity capacity, where the predominant protein is phase Olin, a kind of globulin [2]. The most important world organizations have expressed the constraint of having not enough protein production for an increasing population. The lack of water and farmable land as well as the contribution to climate change due to the cattle industry create an unsustainable situation. For these reasons, is important to find new sources of protein alternatives. This work aimed to develop a food prototype using the hydrolyzed proteins from ñuña beans, mashua (Tropaeolum tuberosum) tuber flour, purple corn (*Zea mays* L.) flour, and pota (*Dosidicus gigas*) by-product meal with a high content of protein and omega-3 acids (~50% EPA + DHA on total fat).

2. Materials and Methods

2.1. Raw Material

Ñuña beans (*Phaseolus vulgaris* L.), mashua (Tropaeolum tuberosum) tuber, and purple corn (*Zea mays L.*) were obtained from the local market in Lima city and pota (*Dosidicus gigas*) by-products were collected by Mi Cautivo de Ayabaca company from Piura Department, Peru. The beans were ground in a disc mill in the Functional Foods Laboratory from the Universidad de Lima-Peru and sieved with 125,250, and 500 µm mesh to obtain four sieved fractions of ñuña bean flour. Pota by-products were washed, dried at 60 °C × 12 h by an infrared dryer (Irconfort IRC D18, Sevilla, Spain), and ground in a disc mill to obtain pota by-product meal. Mashua tuber and purple corn were washed, dried at 40° C, and ground in a food shredder (Grindomix GM200, Restch, Haan, Germany). All the samples were kept at −5 °C in polyethylene bags for later analysis. Alcalase 2.4L was purchased from Sigma Chemical (St. Louis, MO, USA).

2.2. Proximal Composition

The moisture content of the samples was determined at 110 °C to a constant weight. The ash content was determined by the ignition method (550 °C for 72 h). The fat content was determined with hexane for 9 h. The total protein content was determined as % nitrogen × 6.25 using a Kjeldahl analyzer (UDK 139, VELP, Usmate Velate. Italy) with official methods.

2.3. Total Polyphenolic Content

The total phenolic content (TPC) of the samples was determined by the Folin-Ciocalteau method [3]. A 15 mg of sample was dissolved in 4.5 mL of methanol and 2.5 mL of Folin-Ciocalteau reagent 2 N, and the mixture was stirred using a vortex for 1 min. After 5 min, 2.5 mL of sodium carbonate solution (20%) was added, and the mixture was left at 40 °C for 30 min. The mixture was cooled and filtered through Whatman filter paper N° 2 (Whatman International Ltd., Maidstone, UK). The absorbance of the solution was measured at 760 nm using a spectrophotometer (UV 1280 Vis Spectrophotometer, Shimadzu, Kyoto, Japan). Ultrapure water was used as a control blank. The results were expressed as µg of gallic acid equivalent (GAE)/gram (powder). All analyses were done in triplicate and the results are expressed as mean values.

2.4. Antioxidant Activity

A total of 15 mg of samples were resuspended in 4.5 mL of methanol/acetic acid/water (50:8:42, *v*/*v*/*v*), stirred using a vortex for 1 min, and left in a water bath for 20 min at 80 °C [3]. Then, 3.9 mL of 25 ppm 2, 2-diphenyl-1-picrylhydrazyl (DPPH) radical solution (2.5 mg DPPH in 100 mL MeOH) was added, and the samples were left in the dark at 25 °C. The mixture was stirred using a vortex for 1 min and then filtered through Whatman filter paper N° 2. The absorbance of the solution was measured at 517 nm by spectrometry (UV 1280 Vis Spectrophotometer, Shimadzu, Kyoto, Japan). For a control blank, 500 µL of methanol with 3.9 mL of 25 ppm DPPH radical solution was used. All analyses were carried out in triplicate and the results were expressed as mean values.

2.5. Protein Solubility Curve

The sieved fraction of the ñuña bean (red and gray) flour was dissolved in ultrapure water (10% *w*/*v*) and the pH was adjusted to 12 with NaOH 1 N or HCl 1 N for 1 h at room temperature and centrifuged for 15 min at 10,000× *g* rpm [4]. Six aliquots of the supernatant were taken and the pH was adjusted from 2 to 12 with NaOH 1 N or HCl 1 N. They were then centrifuged for 15 min at 10,000× *g* rpm. The supernatant was measured in protein content according to the Lowry method [5]. The absorbance of the solution was measured at 750 nm using a spectrophotometer (UV 1280 Vis Spectrophotometer, Shimadzu, Kyoto, Japan). The results were expressed as % solubilized protein regarding the total protein content.

2.6. Ñuña Protein Extraction

The ñuña bean (red and gray) sieved fraction flour was dissolved in ultrapure water (10% *w*/*v*) and the pH was adjusted to 10 with NaOH 1 N or HCl 1 N for 1 h at room temperature. The mixture was then centrifuged for 10 min at 10,000× *g* rpm. The supernatant obtained was adjusted to the isoelectric point (pI) of ñuña proteins (pH 4) and centrifuged at 10,000 rpm× *g* for 6 min. The precipitate was washed with ultrapure water [4].

2.7. Hydrolysis of Ñuña Protein

The hydrolysis of ñuña beans protein was performed according to Paz et al. [4] with some modifications, using ultrapure water (10% *w*/*v*, g protein/mL water), pH 8 with NaOH 1 N or HCl 1 N, and 800 rpm for 1 h at 50 °C, using a bioreactor (TEC-BIO-FLEX-II, Tecnal, São Paulo, Brazil). Alcalase 2.4L was added at a ratio enzyme/substrate = 0.32 Anson Unit (AU)/g protein. The inactivation of the enzyme was performed at 85 °C for 15 min. The supernatant obtained was spray-dried in a Büchi B-290 (Büchi Labortechnik AG, Flawil, Switzerland) with a nozzle atomization system (0.7 mm nozzle diameter), air flow rate of 55 m^3/h, compressor air pressure of 50 mbar, inlet and outlet air temperatures 180 °C and 90 °C, respectively, and flow rate feed of 55 mL/min. The dried powders collected were stored in opaque hermetic bags at 5 °C for further analysis.

Degree of Hydrolysis

The degree of hydrolysis was calculated by the determination of free amino groups by reaction with 2,4,6-trinitrobenzenesulphonic acid (TNBS) [6]. The absorbance of the supernatant was measured at 340 nm using a spectrophotometer (UV 1280 Vis Spectrophotometer, Shimadzu, Kyoto, Japan). A calibration curve was constructed using L-Leucine (0.10–2.5 mM) solution.

2.8. Development of Food Prototype

The food prototype (porridge powder) was developed by mixing hydrolyzed protein from gray ñuña (*Phaseolus vulgaris* L.) bean with pota (*Dosidicus gigas*) by-product meal, mashua (Tropaeolum tuberosum) tuber flour, and purple corn (*Zea mays* L.) flour. The formulation was kept at −5 °C in a polyethylene bag.

3. Results and Discussions

According to Table 1, the proximate composition was evaluated in regard to their particle size. The gray ñuña bean retained less moisture content (11.46 ± 0.04% to 11.94 ± 0.06%) than red ñuña bean (12.02 ± 0.02% to 12.67 ± 0.02%). The gray ñuña bean had the highest protein content (21.06 ± 2.12% to 30.11 ± 0.82%, fractions between 0–500 µm), followed by the red ñuña bean (18.82 ± 0.37% to 23.9 ± 0.15%, fractions between 0–500 µm). The fat content presented values between 0.43 ± 0.04% to 1.89 ± 0.03% (red ñuña bean) and 0.42 ± 0.03% to 2.26 ± 0.03% (gray ñuña bean). The ash content ranged from 4.04 ± 0.07% to 5.92 ± 0.05% (red ñuña bean) and 3.88 ± 0.09% to 5.37 ± 0.03% (gray ñuña bean), and a high carbohydrate content was discovered (75.32 ± 0% to 75.81 ± 0.03%) for both red and gray ñuña beans.

The results of TPC and DPPH are shown in Table 2. The red ñuña bean was found to have a higher TPC (8589 ± 110 µg GAE/g powder) compared with the gray ñuña bean (3478 ± 117 µg GAE/g powder). The antioxidant activity of the red ñuña bean ranged from 478 ± 17 to 9879 ± 24 µg trolox/g powder (red ñuña bean) and 148 ± 20 to 5539 ± 11 µg trolox/g powder for gray ñuña bean. According to Xu and Diosady [7], phenolic compounds could interact with isolate proteins, which could create a synergetic effect with hydrolysate proteins from ñuña beans.

Table 1. The proximal composition of ñuña (*Phaseolus vulgaris* L.) bean flour.

Particle Size (μm).	Moisture (g/100 g).		Protein (g/100 g).		Lipids (g/100 g).		Ash (g/100 g).		Carbohydrates (g/100 g).	
	Red Ñuña Bean	Gray Ñuña Bean	Red Ñuña Bean	Gray Ñuña Bean	Red Ñuña Bean	Gray Ñuña Bean	Red Ñuña Bean	Gray Ñuña Bean	Red Ñuña Bean	Gray Ñuña Bean
>500	12.02 ± 0.02	11.79 ± 0.02	6.27 ± 0	6.63 ± 0.01	0.43 ± 0.04	0.42 ± 0.03	5.92 ± 0.05	5.37 ± 0.03	75.32 ± 0	75.81 ± 0.03
250–500	12.54 ± 0.03	11.82 ± 0.01	23.44 ± 1.84	21.06 ± 2.12	1.74 ± 0.07	1.24 ± 0.01	4.41 ± 0.02	4.37 ± 0.24	57.86 ± 1.93	61.38 ± 2.25
125–250	12.63 ± 0.03	11.46 ± 0.04	18.82 ± 0.37	30.11 ± 0.82	1.89 ± 0.03	2.26 ± 0.03	4.23 ± 0.01	4.23 ± 0.02	62.44 ± 0.35	51.95 ± 0.74
0–125	12.67 ± 0.02	11.94 ± 0.06	23.9 ± 0.15	26.86 ± 3.23	1.77 ± 0.1	1.44 ± 0.06	4.04 ± 0.07	3.88 ± 0.09	57.6 ± 0.15	55.94 ± 3.25

Results are expressed as means ± SD (n = 2).

Table 2. The total polyphenolic content (TPC) (µg GAE/g powder) and DPPH (µg Trolox/g powder) from ñuña (*Phaseolus vulgaris* L.) bean flour.

Particle Size (µm)	Total Phenolic Content (µg GAE/g Powder)		DPPH (µg Trolox/g Powder)	
	Red Ñuña Bean	Gray Ñuña Bean	Red Ñuña Bean	Gray Ñuña Bean
>500	8589 ± 110	3478 ± 117	9879 ± 24	5539 ± 11
250–500	2252 ± 103	1397 ± 17	3283 ± 22	2309 ± 5
125–250	411 ± 83	350 ± 15	724 ± 12	125 ± 12
0–125	159 ± 16	255 ± 86	478 ± 17	148 ± 20

Results are expressed as the means ± SD (n = 3).

The protein solubility curve from ñuña beans showed an isoelectric point between pH 4 to 4.5. A greater solubility was obtained between pH 8 to 12. For this reason, the protein extraction from ñuña beans was settled at pH 10 with an isoelectric point at pH 4.

The characterization of protein hydrolysates (PH) from ñuña beans is reported in Table 3. The PH from red ñuña bean retained less moisture content (5.38 ± 0.09%). The PH from gray ñuña bean was found to have a higher protein content (69.78 ± 0.82%), high degree of hydrolysis (26.11 ± 0.57%), and ash content (11.62 ± 0.12%). The degree of hydrolysis (26.11 ± 0.57%) of grey ñuña protein hydrolysate was higher than 5%, which indicates that it could be scaled up to industrial steps [8]. For these reasons, the food prototype was formulated with grey ñuña protein hydrolysate. The PH from red ñuña bean showed a higher TPC (55,157 ± 72 µg GAE/g powder) and antioxidant activity (4588 ± 49 µg Trolox/g powder).

Table 3. The characterization of protein hydrolysates from ñuña (*Phaseolus vulgaris* L.) beans.

	Moisture (g/100 g)	Protein (g/100 g)	Degree of Hydrolysis (%)	Ash (g/100 g)	Total Phenolic Content (µg GAE/g)	DPPH (µg Trolox/g)
Red ñuña bean	5.38 ± 0.09	58.42 ± 0.16	15.06 ± 0.15	11.47 ± 0.07	55,157 ± 724	4588 ± 49
Gray ñuña bean	5.96 ± 0.26	69.78 ± 0.82	26.11 ± 0.57	11.62 ± 0.12	12,323 ± 76	740 ± 25

Results are expressed as means ± SD (n = 3).

The purple porridge showed a high protein content (25.47 ± 0.34%), a low percentage of moisture content (7.46 ± 0.09%), and a high number of phenolic compounds (42,429 ± 202 µg GAE/g powder) and antioxidant activity (14,938 ± 25 µg Trolox/g powder), with an acceptable taste and a deep purple color. These results indicate that purple porridge could be an important source of protein and polyphenols for children with malnutrition states.

4. Conclusions

Ñuña beans are a promising source of plant-based protein (around 20%). In addition, the ñuña protein hydrolysates show higher phenolic content (12,323 ± 76 µg GAE/g powder) and antioxidant activity (740 ± 25 µg Trolox/g powder), indicating the developed food prototype would be an important source of protein (25.47 ± 0.34%), polyphenols (42,429 ± 202 µg GAE/g powder), and antioxidant activity (14,938 ± 25 µg Trolox/g powder), all while having an acceptable taste and deep purple color.

Author Contributions: All authors have contributed equally to this manuscript. Conceptualization, N.C., R.A., B.F.G., A.S., M.T., B.G., and M.C.P.-C.; Methodology, N.C., R.A., B.F.G., A.S., B.G., and M.C.P.-C.; Investigation and Data analysis, N.C., R.A., B.F.G., A.S., M.T., P.J., B.G. and M.C.P.-C.; Writing—original draft preparation, N.C., B.F.G., and M.C.P.-C.; Writing—review and editing, N.C., B.F.G., and M.C.P.-C. All authors have read and agreed to the published version of the manuscript.

Funding: This research was funded by by grant Ia ValSe-Food-CYTED (119RT0567) and project SP-2021-01581—Programa Nacional de Investigación en Pesca y Acuicultura (PNIPA), belonging to "the Ministry of Production-Peru", and "Universidad de Lima-Peru".

Institutional Review Board Statement: Not applicable.

Informed Consent Statement: Not applicable.

Data Availability Statement: Data is contained within the manuscript.

Acknowledgments: The authors thank the Instituto de la Grasa-CSIC, Sevilla-España, and Universidad de Lima-Perú.

Conflicts of Interest: The authors declare no conflict of interest.

References

1. Santa Cruz Padilla, A.E.; Vásquez Orrillo, J.L. *Catálogo de Ñuña (Phaseolus Vulgaris L.) Del Banco de Germoplasma Del INIA*; Instituto Nacional de Innovación Agraria, Ed.; Instituto Nacional de Innovación Agraria: Lima, Peru, 2021.
2. Vargas-Salazar, T.A.; Wilkinson, K.A.; Urquiaga Zavaleta, J.M.; Rodríguez-Zevallos, A.R. Physicochemical Charaacterization of Ñuña Bean (*Phaseolus vulgaris* L.) Protein Extract. *Vitae* **2020**, *27*, 1–9. [CrossRef]
3. Luo, Q.; Zhang, J.R.; Li, H.B.; Wu, D.T.; Geng, F.; Corke, H.; Wei, X.L.; Gan, R.Y. Green Extraction of Antioxidant Polyphenols from Green Tea (Camellia Sinensis). *Antioxidants* **2020**, *9*, 785. [CrossRef] [PubMed]
4. Paz, S.M.D.l.; Martinez-Lopez, A.; Villanueva-Lazo, A.; Pedroche, J.; Millan, F.; Millan-Linares, M.C. Identification and Characterization of Novel Antioxidant Protein Hydrolysates from Kiwicha (*Amaranthus caudatus* L.). *Antioxidants* **2021**, *10*, 645. [CrossRef] [PubMed]
5. Lowry, O.H.; Rosebrough, N.J.; Farr, A.L.; Randall, R.J. Protein Measurement with the Folin Phenol Reagent. *J. Biol. Chem.* **1951**, *193*, 265–275. [CrossRef]
6. Adler-Nissen, J. Determination of the Degree of Hydrolysis of Food Protein Hydrolysates by Trinitrobenzenesulfonic Acid. *J. Agric. Food Chem.* **1979**, *27*, 1256–1262. [CrossRef] [PubMed]
7. Xu, L.; Diosady, L.L. Interactions between Canola Proteins and Phenolic Compounds in Aqueous Media. *Food Res. Int.* **2000**, *33*, 725–731. [CrossRef]
8. Bui, X.D.; Vo, C.T.; Bui, V.C.; Pham, T.M.; Bui, T.T.H.; Nguyen-Sy, T.; Nguyen, T.D.P.; Chew, K.W.; Mukatova, M.D.; Show, P.L. Optimization of Production Parameters of Fish Protein Hydrolysate from Sarda Orientalis Black Muscle (by-Product) Using Protease Enzyme. *Clean Technol. Environ. Policy* **2020**, *23*, 31–40. [CrossRef]

Proceeding Paper

Development of Breads Fortified in Calcium and High Protein Content through the Use of Bean Flour and Regional Fruits †

Natalia Bassett [1,2,*], Abigail Gutierrez [2], Elina Acuña [2] and Analia Rossi [2]

1 Centro Interdisciplinario de Investigaciones, Tecnologías y Desarrollo Social Para el NOA (CIITED) Consejo Nacional de Investigaciones Científicas y Técnicas (CONICET), Universidad Nacional de Jujuy, San Salvador de Jujuy 4600, Argentina
2 Instituto Superior de Investigaciones Biológicas (INSIBIO), CONICET, Universidad Nacional de Tucumán, San Miguel de Tucumán 4000, Argentina
* Correspondence: natybassett@gmail.com
† Presented at the IV Conference Ia ValSe-Food CYTED and VII Symposium Chia-Link, La Plata and Jujuy, Argentina, 14–18 November 2022.

Abstract: The World Health Organization recommends the consumption of legumes and considers them good allies to achieve food security and reduce malnutrition worldwide. Bread offers the possibility of incorporating ingredients to improve one's diet without changing their eating habits. The objective of this study was to formulate and elaborate calcium-fortified bread, optimizing nutritional quality using protein supplementation with regionally produced ingredients, which are underutilized for domestic consumption. The variability of the protein quality of wheat flour and its mixtures with bean flour was studied. The proteins, fats, dietary fibre, ashes, and moisture were determined using AOAC methods. Volume, texture, and colour were also evaluated using a Vernier caliper, a TAXT plus Texturometer, and a Colour Quest XE spectrophotometer, respectively. The addition of calcium salts increased hardness, produced lighter crumbs and crust, and did not affect volume. The addition of fruit pulp did modify the colour and volume of the loaves. The moisture, protein, calcium, and sodium content of the baked goods were 42 g, 11.6 g, 443 mg, and 347 mg per 100 g of bread, respectively. A sodium reduction of 30% was obtained with the consequent increase in calcium, both critical nutrients by default. The breads produced are inexpensive and have higher contents of high-quality protein and calcium. Due to their ingredients and their nutritional and textural characteristics, the breads could be incorporated into the diets of vulnerable groups and contribute to the prevention of chronic and/or deficiency diseases. In addition, the use of regional products will encourage local production and therefore support the local economy.

Keywords: added calcium; breads; legumes; fruits; regional

Citation: Bassett, N.; Gutierrez, A.; Acuña, E.; Rossi, A. Development of Breads Fortified in Calcium and High Protein Content through the Use of Bean Flour and Regional Fruits. *Biol. Life Sci. Forum* **2022**, *17*, 19. https://doi.org/10.3390/blsf2022017019

Academic Editors: Norma Sammán, Mabel Cristina Tomás, Loreto Muñoz and Claudia Monika Haros

Published: 2 November 2022

1. Introduction

The world population has grown steadily, and most people now live in urban areas. Technology has evolved at a dizzying pace, while the economy has become increasingly interconnected and globalized. Climate change with the increasing variability of climate and extreme events is affecting agricultural productivity, food production, and natural resources with impacts on food systems and means of rural livelihoods, including a decrease in the number of farmers. All of this has led to major changes in the ways in which food is produced, distributed, and consumed worldwide, and to new food security, nutrition, and health challenges [1] Globally, most markets offer a wide variety of foods and beverages that combine flavour, comfort, and novelty. However, at the same time, there is a wide availability and widespread marketing of many of these products, especially those with a high content in fat, simple sugars, and/or salt. These dietary patterns have an impact on the nutritional status of the Argentine population in general and the NOA population in particular, which has high rates of overweight and obesity that coexist with high rates of malnutrition and low weight, as well as some micronutrient deficits [2,3]. Adding to the epidemic of overweight and obesity, which is the most frequent form of malnutrition and continues to increase steadily in Argentina, is the COVID-19 pandemic [4].

On the one hand, legume flours provide protein, fibre, and micronutrients while reducing fat consumption. They are the fundamental protein and caloric base of human nutrition and the main source of energy and essential amino acids provided in a balanced and nutritious way, especially when they are combined with cereals. Populations in the interior of the country have higher consumptions of dried legumes, compared to the Federal Capital and Greater Buenos Aires. However, in previous studies carried out in the NOA population, little or no consumption of these was observed [2]. The market for legume flour is mainly driven by the health benefits that it provides. When cereal proteins and legumes are combined, the biological value of each of them is improved, making an opportunity to create new foods of high nutritional and organoleptic quality. On the other hand, pitanga (*Eugenia uniflora* L.) is an edible fruit and botanical berry. These native fruits can be an important source of antioxidant and nutraceutical compounds due to the content of calcium, phosphorus, anthocyanins, flavonoids, carotenoids, and vitamin C comparable or superior to others. They are generally consumed fresh, but due to their highly perishable nature, they are also processed into a range of value-added industrialized products such as juices and fruit syrups, among other products [5].

The objective of the present study was to formulate and elaborate a calcium-fortified bread, optimizing nutritional quality using protein supplementation with regionally produced ingredients that are underutilized for domestic consumption. This will provide nutrients for the prevention of chronic and/or deficiency diseases in the populations of Tucumán and Jujuy.

2. Materials and Methods

2.1. Assessment and Selection of the Theoretical Protein Quality of Bread Formulations

The variability of the protein quality of wheat flour and its mixtures with legume flour was studied. Protein quality was quantified based on the amount and profile of essential amino acids (IAAs), as well as the actual ileal digestibility of protein IAAs using the "Digestible indispensable Amino Acid Score" (DIAAS). The value for each IAA in the diet was calculated, and the lowest value was designated as the DIAAS. For this, the computer tool MixProtLUNA.1-2013 created by the working group [6] was used.

2.2. Materials and Elaborated Bread Proximate Analysis

The commercial wheat flour type 000, white bean (*Phaseolus vulgaris*), pitanga (*Eugenia uniflora* L.), and dry instant yeast Calsa® (Buenos Aires, Argentina) were purchased at a local market in Tucumán, Argentina. The integral flour of white beans was made by grinding and sieving them in a No. 35 sieve. For the elaboration of bread, wheat flour was partially replaced with legume flour (white bean flour) and the mixture of calcium and sodium salts was incorporated, which was obtained in previous works as the optimal mixture for obtaining a dough for bread [7]. The amount of pitanga (*Eugenia uniflora* L.) pulp to be added was studied, which did not modify the textural characteristics of the baked product. The salt substitute mix was made with NaCl and $CaCO_3$ from (Cicarelli® Santa Fé, Argentina). The bread making was carried out in an Atma HP4031E Bread Oven. The elaborated breads were kept in plastic bags of 100 g each in a dry place at room temperature. Mass yield, product yield, and specific volume calculations were performed. The breads were analysed for protein, fibre, fat, ash, and moisture using AOAC [8]. The available carbohydrate content (g/100 g) was calculated by subtracting the contents of moisture, fat, dietary fibre, ash, and protein from 100%. Calcium and sodium content were determined using an atomic absorption spectrometer Perkin Elmer PinAAcle 900T (Akron, OH, USA).

2.3. Determination of Texture and Colour

The textural characteristics of the breads were determined using the TA.XT Plus Texturometer (Stable Micro Systems Ltd., Surrey, UK) containing a 50 kg maximum load cell. The samples were analysed after 10 h of cooking. An average of three measurements was made.

Colour analyses were performed with a Colour Quest XE spectrophotometer (Hunter Lab, Reston, VA, USA) with D65 illuminate, a 0° standard observer, and a 2.5 cm port/viewing area. The determinations were made through the crumb and the upper and lower crusts of the breads ($n = 6$) and was expressed with the L*, a*, and b* parameters.

2.4. Statistical Analysis

Analyses were performed using the IBM SPSS Advanced Statistics 23.0 (IBM Software Group, Chicago, IL, USA). The significant difference between the means was evaluated by Tukey's test ($p < 0.05$) using analysis of variance (ANOVA). All determinations were made in triplicate using different batches of bread samples.

3. Results and Discussion

3.1. Theoretical Protein Quality Assessment

The protein content of the commercial wheat flour used was 10 g/100 g and obtained a DIAAS of 47%, while white bean flour with a protein content of 27.3 g/100 g obtained a DIAAS of 65%. Therefore, the formulation of mixtures of cereals and legumes improves the amino acid balance and translates into a higher value in the quality of the protein compared to that of each one separately. Theoretical combinations were made from additions of 10% to 90% until finding the most suitable proteins for each flour component of the mixture and achieving a product of greater nutritional value using the MixProtLUNA computer program. The results are shown in Table 1 where it is observed that the DIAASs range from 47 to 89%. Theoretical recipes were formulated with the different percentages and then made into breads, each of them with a flour/water ratio of 100/65 and with a content of 1% NaCl salt, 1% calcium salt, and 1% yeast. The bread that finally resulted with adequate baking characteristics was the one that had a 20/80 combination of bean proteins with commercial wheat flour DIAAS: 62.

Table 1. Digestible indispensable amino acid score (DIAAS) in different blends in percentage of wheat flour protein with white bean flour protein.

FOOD/VARIETY/ORIGIN/TYPE White bean flour (*P. Vulgaris*) P: 27.3, D: 65; AALim: SAA Wheat flour (*Triticum* spp.) P:10, D:47; AALim: Lys										
HPB/HT RATIOS										
0–100	10–90	20–80	30–70	40–60	50–50	60–40	70–30	80–20	90–10	100–0
DIAAS %										
47	55	62	69	75	80	84	88	89	79	65

HPB = White bean flour; HT= Wheat flour; DIAAS = Digestible Indispensable Amino Acid Score. AALim: limiting amino acid. Lys: lysine; SAA: sulfur amino acids (Methionine + cysteine) P = Protein g/100 g food.

3.2. Materials and Elaborate Bread Proximate Analysis

Bread is widely consumed; therefore, it could be used as a suitable vehicle to supply high-quality nutrients to consumers without changing their eating habits. A good dough yield of 179% and a bread yield of 90% were obtained (Figure 1). Table 2 shows the chemical composition of the elaborated breads. It can be seen that there were significant differences in all the parameters analysed except for the ash content. Moisture varied from 40.7 to 43.9, being higher in breads with additions of bean flour, pitanga pulp, and calcium salts. Regarding the content of protein, lipids, and dietary fibre, a higher value was obtained in the breads with the addition of bean flour. Regarding the calcium content, it was observed that the content of calcium salts increased in fortified breads with the addition of white bean flour, as expected. Calcium is a nutrient that has limited food sources. Dairy products (milks, yogurts, and cheeses) are the main source of dietary calcium. Despite Argentina being placed among the countries with an important dairy industry, the food availability of calcium is insufficient to cover the population requirements [9]. A serving size of about 40 g of fortified bread already covers 20% of the IDR for an adult (average body weight 70 kg with moderate physical activity, taking into account that the recommendation for calcium intake per day is 1000 mg). The substitution has an inverse effect with respect to sodium intake. Sodium was higher in breads made only with white bean flour, despite this; all baked goods are in accordance with Law 26905 of the Honourable Congress of the Argentine nation's maximum sodium values published in the official bulletin of 2013 number: 32786. Other authors who produced food to cover calcium deficiencies showed similar values, although many of them did not show a sodium decrease and, in some cases, they added sugars. In the future, it would be advisable to carry out acceptability surveys in especially vulnerable groups and populations in which calcium consumption is deficient, as would be case of pregnant and hypertensive people.

Figure 1. Appearance of pitanga fruit, white beans, and elaborated bread.

Table 2. Proximal compositions (g/100g fresh weight) of different types elaborated breads.

	White Bread	White Calcium Bread **	White Bean Bread	White Bean Calcium Bread **	White Bean Pitanga Bread	White Bean Calcium Pitanga **	p Value
Energy (kcal)	220 ± 2 [a,b]	221 ± 3 [a,b]	229 ± 3 [b]	223 ± 1 [a,b]	225 ± 5 [a,b]	216 ± 4 [a]	0.026
Moisture (g)	41.0 ± 0.6 [a,b]	40.7 ± 0.7 [a]	40.8 ± 0.4 [a]	42.5 ± 0.4 [b,c]	41.5 ± 0.5 [a,b]	$42{,}1 \pm 0.7$ [c]	0.001
CHa (g)	43.4 [b]	43.5 [d]	35.2 [a]	34.4 [a]	35.8 [a]	33.5 [a]	0.000
Protein (g)	9.8 ± 0.0 [a]	10.2 ± 0.3 [b]	11.3 ± 0.3 [b,c]	11.2 ± 0.4 [b,c]	11.6 ± 0.6 [c]	11.4 ± 0.3 [b,c]	0.004
Lipid (g)	0.8 ± 0.2 [a]	0.7 ± 0.1 [a]	2.7 ± 0.5 [b]	2.5 ± 0.7[b]	2.0 ± 0.4 [b]	2,72 +/- 0,13 [b]	0.000
Ash (g)	2.4 ± 0.2	2.0 0.1	2.0 ± 0.4	1.8 ± 0.2	2.0 ± 0.2	2.2 ± 0.3	0.274
DF (g)	$2{,}6 \pm 0.5$ [a]	$3{,}0 \pm 0.3$ [a]	$5{,}9 \pm 0.7$ [b]	$5{,}6 \pm 0.4$ [b,]	$7{,}2 \pm 0.1$ [c]	$7{,}34 \pm 0.3$ [c]	0.000
Ca (mg) #	25.7 ± 0.3 [a] (1%)	435.0 ± 25.1 [b] (17%)	29.2 ± 3.0 [a] (1%)	488.3 ± 22.6 [b] (20%)	34.5 ± 3.4 [a] (1%)	439.3 ± 38.1 [b] (18%)	0.000
Na (mg)	349.5 ± 14.8 [a]	331.4 ± 16.4 [a]	864.1 ± 11.1 [b]	326.7 ± 18.7 [a]	835.8 ± 16.1 [b]	346.8 ± 18.4 [a]	0.000

Mean ± SD; $n = 3$. CHa Carbohydrates available calculated by difference (100-Moisture-Protein-Lipids-Dietary fibre-Ashes). DF: dietary fibre; Ca: calcium (**); Added Ca salts: 2.5 g $CaCO_3$ + 2.5 g $CaCl_2$/500 g of flour. (%) Percentage of coverage of daily intake per 40 g portion of bread # Reference values: IDR: 1000 mg/d according to Ross et al, [10] Different letter in the same line indicated significant difference $p < 0.05$ (Tukey's test).

3.3. Determination of Specific Volume, Texture, and Colour in Elaborated Breads

Table 3 shows the colour measured in the breads made. The results showed significant differences in the four parameters evaluated. The addition of calcium salts increased hardness, produced lighter crumbs and crusts, and did not affect volume with respect to the common white bread. The addition of fruit pulp did modify the colour and specific volume (2.1 mL/g) of the loaves (Table 4). The evaluation of the texture parameters of the breads with legumes, pitanga pulp, and calcium showed significant differences in hardness and chewiness in comparison to the control bread (without added salts or beans), obtaining softer and smoother crumbs in unfortified breads. With the development of these fortified breads, we intended to achieve the revaluation of bean varieties and regional fruits, highlighting their nutritional importance and the agro-industrial potential that they present according to their genetic material. Due to its wide consumption, bread can be used as a suitable vehicle to supply high-quality nutrients to consumers. Therefore, the challenge was to develop a healthy bread that incorporates the consumption of these underutilized products into the daily life of the population. The future research will focus particularly on the acceptability of the bread with white bean flour, calcium, and fruit added.

Table 3. Colour of the crusts and crumbs of breads made with mixtures of wheat flour, beans, and pitanga pulp with and without added calcium salts.

		White Bread	White Calcium Bread	White Bean Bread	White Bean Calcium Bread	White Bean Pitanga Bread	White Bean Calcium Pitanga Bread	p Value
Upper crust	L*	74.8 ± 0.5 d	72.92 ± 0.9 d	65.97 ± 1.10 b, c	61.12 ± 1.37 a	63.99 ± 2.08 b	66.21 ± 0.92 c	0.000
	a*	4.1 ± 0.5 a	4.17 ± 0.39 a	6.76 ± 0.70 b	11.63 ± 1.04 c	6.97 ± 2.89 b	4.22 ± 0.25 a	0.000
	b*	24.4 ± 0.7 b	23.39 ± 0.80 b	28.76 ± 1.06 c	36.28 ± 0.66 d	23.91 ± 1.07 b	20.24 ± 1.00 a	0.000
Crumb	L*	65.8 ± 2.3 b	65.98 ± 0.89 b	70.36 ± 0.82 c	68.92 ± 0.90 b, c	59.07 ± 1.85 a	61.1 ± 0.6 a	0.000
	a*	1.8 ± 0.1 a, b	1.74 ± 0.19 a	2.61 ± 0.16 c	3.03 ± 0.12 d	5.48 ± 0.19 b	4.7 ± 0.5 a, b	0.000
	b*	22.7 ± 0.6 c, d	22.02 ± 0.24 c	22.98 ± 0.64 d	24.20 ± 0.28 e	19.53 ± 0.50 b	18.1 ± 0.6 a	0.000
Lower crust	L*	57.1 ± 1.9 c	62.9 ± 0.7 d	50.9 ± 1.7 b	41.1 ± 2.5 a	55.5 ± 1.5 c	60.9 ± 0.3 d	0.000
	a*	13.2 ± 0.7 b	10.0 ± 0.5 a	15.1 ± 0.6 c	18.9 ± 0.7 d	12.6 ± 0.3 b	10.7 ± 0.5 a	0.000
	b*	34.3 ± 0.8 b	32.8 ± 12 b	38.7 ± 1.2 c	45.3 ± 1.9 d	27.2 ± 1.2 d	26.4 ± 0.3 a	0.000

Data shown are mean values of triplicate analyses ± standard deviation from each type of bread. L*, a*, and b* parameters of colour. Values with different letter in the same line indicated significant difference $p < 0.05$ (Tukey's test).

Table 4. Volume and textural profile of breads made with mixtures of wheat flour, beans, and pitanga pulp with and without added calcium salts.

	White Bread	White Calcium Bread	White Bean Bread	White Bean Calcium Bread	White Bean Pitanga Bread	White Bean Calcium Pitanga Bread	p Value
Volume (mL/g)	2.8 b	2.8 b	2.5 a, b	2.5 a, b	2.1 a	2.0 a	0.005
Hardness (g)	1208.5 a	1578.0 b	1995.3 c	2807.4 e	1026.1 a	2374.9 d	0.000
Adhesiveness	−0.118	−0.147	−0.265	−0.208	0.000	−0.151	
Springiness	0.914	0.885	0.849	0.843	0.943	0.943	
Cohesiveness	0.813	0.782	0.802	0.750	0.819	0.836	
Gumminess	980.6 a, b	1229.7 b	1597.3 c	2108.2 d	840.9 a	1985.6 d	0.000
Chewiness	896.8 a	1088.1 a, b	1355.8 a, b, c	1781.3 d	1219.4 a, b, c	1537.1 c, d	0.001
Resilience	0.492	0.468	0.469	0.430	0.513	0.536	

Data shown are mean values of triplicate analyses from each type of bread. Values with different letter in the same line are indicated significant difference $p < 0.05$ (Tukey's test).

4. Conclusions

The breads elaborated in this study are inexpensive to produce, made from low-cost crops of regional importance, good quality, and higher contents of protein and calcium. A 40 g serving provides 18% of the RDI for calcium, and due to its ingredients and its nutritional and textural characteristics, it is suitable for inclusion in the diets of vulnerable groups. They will also contribute to the prevention of chronic and/or deficiency diseases. In addition, the use of raw materials of regional origin will encourage local production and therefore support the local economy.

Author Contributions: Conceptualization, N.B. and A.R.; methodology, N.B and E.A.; software, N.B.; validation, N.B., E.A. and A.R.; formal analysis, N.B., E.A., A.G.; investigation, N.B.; resources, A.R.; data curation, N.B.; writing—original draft preparation, N.B.; writing—review and editing, N.B.; visualization, N.B., A.G., E.A. and A.R; supervision, A.R.; project administration, A.R.; funding acquisition, A.R. All authors have read and agreed to the published version of the manuscript.

Funding: This work was supported by grant Ia ValSe-Food-CYTED (119RT0567) and INSIBIO, CIITED-CONICET–UNJu.

Institutional Review Board Statement: Not applicable.

Informed Consent Statement: Not applicable.

Data Availability Statement: The data presented in this study are available in figures and tables within this article.

Conflicts of Interest: The authors declare no conflict of interest.

References

1. Food and Agriculture Organization of the United Nations (FAO); The International Fund for Agricultural Development (IFAD); The United Nations Children's Fund (UNICEF); The World Food Programme (WFP); The World Health Organization (WHO). The State of Food Security and Nutrition in the World 2019: Safeguarding against Economic Slowdowns and Downturns. Rome, Italy, 2019; p. 572. Available online: http://www.fao.org/3/CA3129EN/CA3129EN.pdf (accessed on 15 July 2021).
2. Bassett, M.N.; Gimenez, M.A.; Romaguera, D.; Sammán, N. Nutritional status and food intake of populations from high altitude regions of the Northwest of Argentina. *Arch. Latinoam. Nutr.* **2013**, *63*, 114–124. [PubMed]
3. ENNyS2. Encuesta Nacional de Nutrición y Salud. Informe técnico Septiembre 2019. Ministerio de Salud. Buenos Aires; Argentina. 2019. Available online: http://www.msal.gob.ar/images/stories/bes/graficos/0000001565cnt-ennys2_resumen-ejecutivo-2019.pdf (accessed on 15 July 2021).
4. Bonvecchio, A.; Bernal, J.; Herrera Cuenca, M.; Aldana, M.F.; Gutiérrez, M.; Irizarry, L.; Lay Mendivil, L.; López Bautista, F.; López Reyes, M.; Mata, C.; et al. Recomendaciones de micronutrientes para grupos vulnerables en contexto de desnutrición, durante la pandemia de COVID-19 en Latinoamérica. *Arch. Latinoam. Nutr.* **2019**, *69*, 259–273. [CrossRef]
5. Franzon, R.C.; Carpenedo, S.; Viñoly, M.D.; Raseira, M.d.C.B. Pitanga—*Eugenia uniflora* L. In *Exotic Fruits*; Rodrigues, S., de Oliveira Silva, E., de Brito, E.S., Eds.; Academic Press: Cambridge, MA, USA, 2018; pp. 333–338. [CrossRef]
6. Bassett, M.N.; Espejo, L.M.; Sammán, N.C. Estudio de la calidad de diferentes mezclas de harinas proteicas aplicando el software MixProtLUNA. In Proceedings of the V Congreso Internacional de Ciencia y Tecnología de los Alimentos, Córdoba, Argentina, 17–19 November 2014; pp. 3–10. (In Spanish).
7. Bassett, M.N.; Pérez-Palacios, T.; Cipriano, I.; Cardoso, P.; Ferreira, I.M.; Samman, N.; Pinho, O. Development of bread with NaCl reduction and calcium fortification: Study of its quality characteristics. *J. Food Qual.* **2014**, *37*, 107–116. [CrossRef]
8. AOAC. Association of Official Analytical Chemists. Methods of Analysis (AOAC). Available online: http://www.aoac.org/ (accessed on 1 March 2021).
9. Aballay, L.R.; Eynard, A.R.; Díaz, M.d.P.; Navarro, A.; Muñoz, S.E. Overweight and obesity: A review of their relationship to metabolic syndrome, cardiovascular disease, and cancer in South America. *Nutr. Rev.* **2013**, *71*, 168–179. [CrossRef] [PubMed]
10. Ross, A.C.; Manson, J.E.; Abrams, S.A.; Aloia, J.F.; Brannon, P.M.; Clinton, S.K.; Durazo-Arvizu, R.A.; Gallagher, J.C.; Gallo, R.L.; Jones, G.; et al. The 2011 report on dietary reference intakes for calcium and vitamin D from the Institute of Medicine: What clinicians need to know. *J. Clin. Endocrinol. Metab.* **2011**, *96*, 53–58. [CrossRef] [PubMed]

Proceeding Paper

Immunonutritional Benefits of *Chenopodium quinoa*'s Ingredients Preventing Obesity-Derived Metabolic Imbalances †

Helena Marcos Pasero [1], Adrianna Bojarczuk [2], Claudia Monika Haros [3] and Jose Moisés Laparra Llopis [4,*]

1 Food Science and Health Group, Valencian International University (VIU), 46021 Valencia, Spain
2 Institute of Agricultural and Food Biotechnology, 02-532 Warsaw, Poland
3 Instituto de Agroquímica y Tecnología de Alimentos (IATA), Consejo Superior de Investigaciones Científicas (CSIC), 46980 Valencia, Spain
4 Molecular Immunonutrition Group, Madrid Institute for Advanced Studies in Food (IMDEA-Food), 28049 Madrid, Spain
* Correspondence: moises.laparra@imdea.org
† Presented at the IV Conference Ia ValSe-Food CYTED and VII Symposium Chia-Link, La Plata and Jujuy, Argentina, 14–18 November 2022.

Abstract: Over 1.6 billion people (aged 15 years and above) worldwide are currently either overweight or obese, and this number is predicted to increase to 2.3 billion by 2050 (WHO). Excessive or impaired energy storage in the liver incurs a high risk of liver dysfunction and development of obesity, lipodystrophy, or cachexia, and impairs organismal homeostasis. Chenopodium quinoa seeds constitute a good source of immunonutritional compounds, enabling the selective functional differentiation and function of intrahepatic monocyte-derived macrophages. The latter play a key role in controlling adiposity associated with innate lymphoid cells (ILCs), which determine the induction of diet-induced obesity (DIO). Herein, two immune-conditioned mouse models—Rag2$^{-/-}$ and Rag2$^{-/-}$IL2$^{-/-}$—were used to examine the influences preventing DIO with a protein-rich fraction (PRF) and oil obtained from *C. quinoa* seeds. Variations in myeloid cells and precursors of ILCs were evaluated by FACS analyses as well as the hepatosomatic index to estimate liver inflammation. Only the administration of *C. quinoa* PRF prevented alterations in the liver/body weight ratio, both in animals carrying ILCs (i.e., Rag2$^{-/-}$) and not (Rag2$^{-/-}$IL2$^{-/-}$). These effects were associated with significantly decreased variations in the hepatic triglyceride content. FACS revealed that PRF from *C. quinoa* favors the hepatic infiltration of myeloid and enables the selective functional differentiation and function of intrahepatic monocyte-derived macrophages, preserving tissue integrity and function.

Keywords: obesity; macrophages; serine-type protease inhibitors; quinoa; food

Citation: Marcos Pasero, H.; Bojarczuk, A.; Haros, C.M.; Laparra Llopis, J.M. Immunonutritional Benefits of *Chenopodium quinoa*'s Ingredients Preventing Obesity-Derived Metabolic Imbalances. *Biol. Life Sci. Forum* **2022**, *17*, 20. https://doi.org/10.3390/blsf2022017020

Academic Editors: Norma Sammán, Mabel Cristina Tomás and Loreto Muñoz

Published: 3 November 2022

1. Introduction

Over 1.6 billion people (aged 15 years and above) worldwide are currently either overweight or obese, and this number is predicted to increase to 2.3 billion by 2050 (WHO). Recent research has highlighted the key role of innate and adaptive lymphocytes to operate sequentially and in distinct ways during normal development to establish tissue lipid homeostasis [1]. Moreover, innate lymphoid cells (ILCs) have been identified as determinants in the induction of diet-induced obesity [2]. Non-alcoholic fatty liver disease (NAFLD) has become the most common liver pathology worldwide, affecting an estimated 15–30% of most populations due to dramatic increases in risk factors such as obesity, sedentary lifestyle, and altered food supplies and preferences. Between 10% and 20% of subjects with NAFLD will have a severe variant of non-alcoholic steatohepatitis (NASH), where the fatty liver has progressed to massive hepatocyte apoptosis (mitochondrial dysfunction and lipoapoptosis due to an excess of free fatty acids and uncontrolled oxidative processes), hepatic inflammation, and the development of liver fibrosis. Fibrosis progresses to cirrhosis in 10–20% of patients with persistent NASH, with high liver-related morbidity and mortality, part of which is due to the development of hepatocellular carcinoma (HCC) [3]. Here, macrophage-dependent mechanisms appear to control energy storage into the liver [4].

Previous research evidenced the potential role of ingredients from *C. quinoa* to modulate insulin resistance as well as hepatic triglyceride accumulation [5]. Immunonutritional bioactive compounds from *C. quinoa* positively preserved alterations in peripheral myeloid populations [6]. However, to the best of our knowledge, there is no information about the influence of *C. quinoa* on the maturation of ILCs and their association with the selective functional differentiation of monocyte-derived macrophage.

In this respect, this study explored the impact of the administration of *C. quinoa*'s ingredients in the differentiation of innate immune effectors key in liver fat mobilization and energy storage.

2. Materials and Methods

2.1. Ingredients

C. quinoa germs were obtained by cold pressing and from oil by wet milling [7]. A protein-rich fraction from the germ was obtained by vigorous vortexing with phosphate-buffered solution (PBS); 1 h/25 °C.

2.2. Experimental Model and Subject

All animal experiments were conducted in accordance with the guidelines of the Institutional Animal Care and Use Committee of the Spanish High Research Council of Research (CSIC) and regional government of Madrid (Spain) (Proex 298.0/21). Mice used in this study were on a C57BL/6 background and were maintained in the CIB-CSIC Center under pathogen-free conditions. Rag2$^{-/-}$ and Rag2$^{-/-}$IL2$^{-/-}$ mice were kindly donated by Dr. Balbino Alarcón (CBM-CSIC). Mice (6 weeks old) were placed on a 42 kcal% HFD (IN93G mod, irradiated, Ssniff spezialdiäten gmbh) for 15 days and received the protein-rich fraction (PRF) (10 µg) and oil (0.8 µL) from *C. quinoa* 3 times per week. The diet was exchanged twice a week, and body weights were measured every 2 days. Liver weight as a percentage of the body weight (hepatosomatic index) was calculated. Liver samples were collected and stored at −80 °C until analysis.

2.3. Cell Preparation for Flow Cytometry and the Gating Strategy

Cells in the liver tissue were prepared as previously described, with a few modifications [2]. Briefly, the liver was cut into 1 cm pieces and stirred in a 100 mL conical flask containing 5 mL of 1 mM EDTA in PBS with a stir bar at 37 °C and 500 rpm for 20 min. The pieces were shaken vigorously in 20 mL PBS, minced into 1–2 mm pieces with scissors, and stirred in 10 mL of 2% (*v*/*v*) FCS in RPMI medium containing Trypsin-EDTA solution, 0.25%, sterile-filtered, suitable for cell culture, 2.5 g porcine trypsin and 0.2 g EDTA, 4Na per liter of Hanks' Balanced Salt Solution (HBSS) with phenol red with a stir bar at 37 °C and 500 rpm for 60 min. Finally, samples were filtered through a 40 µm strainer into a new 50 mL tube. The tubes were centrifuged at room temperature and 500× *g* for 10 min, and the pellets were then used for subsequent analyses. Prepared cells were suspended in 2% FCS in HBSS containing 0.02% NaN_3, and propidium iodide was added to gate out dead cells immediately before analysis. Flow cytometry was performed using FACS SORP apparatus (BD Biosciences, Franklin Lakes, NJ, USA). Data were analyzed using FlowJo Software (FlowJo LLC, Ashland, OR, USA). Based on the surface expression levels of NKp46, CD56, and killer cell lectin-like receptor G1 (KLRG1), four subsets of immature ILCs could be distinguished [8]: (i) one subset consisted of CD117$^+$NKp46$^-$CD56$^-$KLRG1$^-$ cells that could be multipotential; (ii) two immature CD117$^+$NKp46$^+$ and CD117$^+$NKp46$^+$CD56$^+$ subsets were biased to mature into group 3 ILCs (ILC3s); and (iii) an immature CD117$^+$ KLRG1$^+$ subset was biased to mature into group 2 ILCs (ILC2s). For analysis of macrophages, PI-CD68$^+$F4/80$^+$ cells were identified as M1 macrophages.

2.4. Statistical Analyses

Statistical analysis between the different groups of treatment within the same experimental model was conducted using one-way analysis of variance (ANOVA) and the Kruskal–Wallis post hoc test by ranks. Analyses were performed with Statgraphics Centurion XVI software (Statgraphics Technologies, Inc., The Plains, VA, USA), and significance was established at $p < 0.05$ for all comparisons.

3. Results

Animals administered with PRF displayed downward trends in body weight gain (Figure 1A–C) rates in relation to controls. The administration of *C. quinoa*'s ingredients prevented increased values in the hepatosomatic index (Figure 1D) in both animal models.

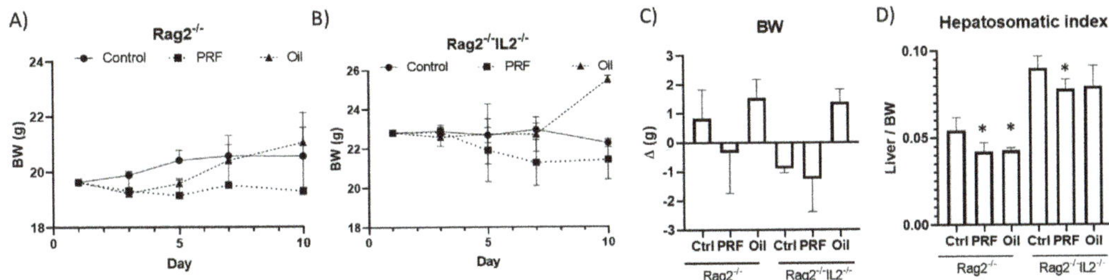

Figure 1. Morphometric measurements. Daily (**A**,**B**) and total (**C**) body weight gain (BW), and hepatosomatic index (**D**) of Rag2$^{-/-}$ and Rag2$^{-/-}$Il2$^{-/-}$ mice fed a high-fat diet and administered with a protein-rich fraction (PRF) or oil from *C. quinoa*. Results are expressed as the mean ± standard error (SEM) (n = 6). * Indicates statistical differences in relation to the respective controls.

Animals receiving either PRF or oil exhibited increased proportions of infiltrated monocytes (Figure 2A) to a different extent. These values were associated with mirage trends in the proportion of immature ILC2s (Figure 2B). Notably, only administration of PRF enabled decreased contents of hepatic triglycerides (Figure 2C), whereas both PRF and oil promoted upward amounts of saturated triglycerides, although only in animals carrying ILCs (Rag2$^{-/-}$).

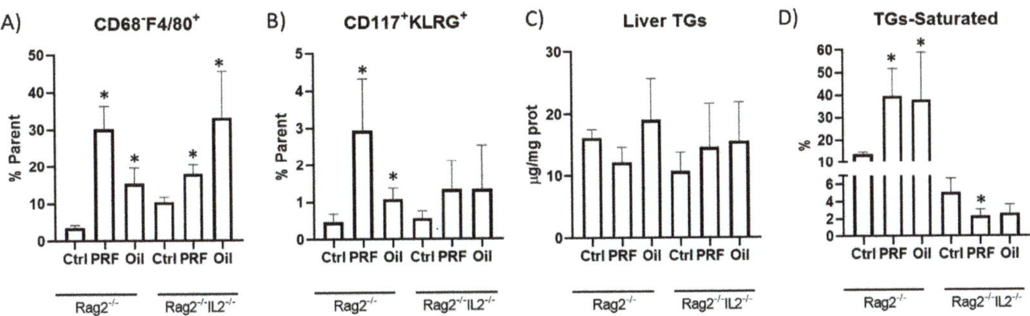

Figure 2. Immunonutritional measurements. Changes in hepatic infiltrated myeloid (**A**) and immature innate lymphoid cells type 2 (i-ILC2s) (**B**), hepatic triglycerides (TGs) (**C**) and saturated TGs (**D**) in Rag2$^{-/-}$ and Rag2$^{-/-}$Il2$^{-/-}$ mice fed a high-fat diet and administered with a protein-rich fraction (PRF) or oil from *C. quinoa*. Results are expressed as the mean ± standard error (SEM) (n = 6). * Indicates statistical differences in relation to the respective controls.

4. Discussion

This study investigated the influence of the independent administration of quinoa's ingredients on hepatic homeostasis in animals fed an HFD. The administration of *C. quinoa*'s ingredients boosted immunonutritional hepatic features, which was reflected in decreased hepatosomatic index values. Here, triglyceride breakdown from hepatic tissue seems to regulate inflammation. Monocyte-derived macrophages have been shown to play key roles coupling energy intake with fat accumulation [4]. Thus, by promoting hepatic TGs mobilization, infiltrated macrophages appear to regulate the systemic availability of lipids in animals under an HFD.

ILC2s and ILC3s are involved in the progression of NAFLD. The influence of PRF, enabling the proliferation of immature ILC2s in an apparent independent fashion from inflammation, could have important consequences in vivo, limiting obesity and insulin resistance and controlling hepatic metabolic homeostasis. ILC2s also interact with other immune cells through newly identified pathways in the adipose tissue. Overall, it can be hypothesized that beneficial effects of the IL-33/ILC2s axis during early stages of diet-induced obesity could result in a potential deleterious role of endogenous IL-33 in late stages of the disease.

5. Conclusions

C. quinoa's ingredients display a different potential to induce relevant innate immune myeloid and lymphoid populations. These changes evidence the clear impact that lipid homeostasis has on diet-induced low-grade inflammation in obesity.

Author Contributions: Conceptualization, J.M.L.L.; methodology, J.M.L.L.; validation, J.M.L.L., H.M.P., C.M.H.; formal analysis, A.B., H.M.P.; investigation, all authors; resources, J.M.L.L., and C.M.H.; data curation, J.M.L.L.; writing—original draft preparation, H.M.P., A.B., and J.M.L.L.; writing—review and editing, J.M.L.L. and C.M.H.; project administration, J.M.L.L. and C.M.H.; funding acquisition, J.M.L.L. and C.M.H. All authors have read and agreed to the published version of the manuscript.

Funding: This work was financially supported by grants Food4ImNut (PRP_PID-2019-107650RBC22, Ministry of Science, Innovation and Universities—MICIU) and CYTED (La ValSe-Food, 119RT0S67); by Ministerio de Ciencia e Innovación, Spain (PID2019-110183RB-C21).

Institutional Review Board Statement: Animal experiments were carried out in strict accordance-with the recommendations in the Guide for the Care and Use of Laboratory Animals of CSIC (Consejo Superior de Investigaciones Científicas), and the protocol was approved by their ethic committee and the regional government (Ethic code, Proex 220/17, approved 21/01/2018).

Informed Consent Statement: Not applicable.

Data Availability Statement: The data that support the findings of this study are available from the corresponding author upon reasonable request.

Acknowledgments: This work was supported by grant Ia ValSe-Food-CYTED (119RT0567) and PID2019-107650RB-C22.

Conflicts of Interest: The authors declare no conflict of interest.

References

1. Mao, K.; Baptista, A.P.; Tamoutounour, S.; Zhuang, L.; Bouladoux, N.; Martins, A.J.; Huang, Y.; Gerner, M.Y.; Belkaid, Y.; Germain, R.N. Innate and adaptive lymphocytes sequentially shape the gut microbiota and lipid metabolism. *Nature* **2018**, *554*, 255–259. [CrossRef]
2. Sasaki, T.; Moro, K.; Kubota, T.; Kubota, N.; Kato, T.; Ohno, H.; Nakae, S.; Saito, H.; Koyasu, S. Innate Lymphoid Cells in the Induction of Obesity. *Cell Rep.* **2019**, *28*, 202–217. [CrossRef] [PubMed]
3. Schuppan, D.; Schattenberg, J.M. Non-alcoholic steatohepatitis: Pathogenesis and novel therapeutic approaches. *J. Gastroenterol. Hepatol.* **2013**, *28*, 68–76. [CrossRef]
4. Cox, N.; Crozet, L.; Holtman, I.R.; Loyher, P.L.; Lazarov, T.; White, J.B.; Mass, E.; Stanley, E.R.; Elemento, O.; Glass, C.K.; et al. Diet-regulated production of PDGFcc by macrophages controls energy storage. *Science* **2021**, *373*, eabe9383. [CrossRef] [PubMed]

5. Selma-Gracia, R.; Haros, C.M.; Laparra Llopis, J.M. Inclusion of Salvia hispanica L. and Chenopodium quinoa into bread formulations improves metabolic imbalances derived from a high-fat intake in hyperglycaemic mice. *Food Funct.* **2020**, *11*, 7994–8002. [CrossRef]
6. Selma-Gracia, R.; Megušar, P.; Haros, C.M.; Laparra, J.M. Immunonutritional Bioactives from Chenopodium quinoa and Salvia hispanica L. Flour Positively Modulate Insulin Resistance and Preserve Alterations in Peripheral Myeloid Population. *Nutrients* **2021**, *13*, 1537. [CrossRef] [PubMed]
7. Ballester-Sánchez, J.; Gil, J.V.; Fernández-Espinar, M.T.; Haros, C.M. Quinoa wet-milling: Effect of steeping conditions on starch recovery and quality. *Food Hydrocoll.* **2019**, *89*, 837–843. [CrossRef]
8. Bal, S.M.; Golebski, K.; Spits, G. Plasticity of innate lymphoid cell subsets. *Nat. Rev. Immunol.* **2020**, *20*, 552–565. [CrossRef] [PubMed]

Proceeding Paper

Chaco Prickly Pear (*Cereus forbesii* Otto ex C.F. Först): An Ancient Source of Antioxidants and Dietary Fiber in the Diet of Indigenous Populations and Its Potential Application as an Ingredient in Derived Products †

Jennifer López, Rocío Villalba, Silvia Caballero, Eva Coronel, Patricia Piris, Adeline Friesen and Laura Mereles *

Facultad de Ciencias Químicas, Universidad Nacional de Asunción, San Lorenzo P.O. Box 1055, Paraguay
* Correspondence: laurameereles@qui.una.py

† Presented at the IV Conference Ia ValSe-Food CYTED and VII Symposium Chia-Link, La Plata and Jujuy, Argentina, 14–18 November 2022.

Abstract: In regions with a majority population of people belonging to indigenous peoples, the solutions to nutritional challenges such as overweight and obesity can go through the implementation of public policies that encourage the use of local and ancestral crops, which would also entail the protection of food traditions. However, these foods can also be added to diets as processed products with high nutritional value. This work describes the centesimal composition and antioxidant potential of a wild prickly pear (*Cereus forbesii* Otto ex C.F. Först) from the Paraguayan Chaco, ancestral food of indigenous peoples, and the jam of this native fruit, a derived product, with the aim of making its nutritional potential known, its potential application in feeding programs and its incorporation in minimally processed foods. These foods mainly show an interesting contribution of micronutrients, soluble sugars, dietary fiber, and antioxidants as anthocyanins with an attractive color, that can replace critical nutrients such as artificial additives and excess sugars in the diet of the regional population. Knowledge of the nutritional and technological properties of regional foods will help strengthen and develop national and regional policies and programs for the development and promotion of local and indigenous products, within the framework of food safety.

Keywords: autochthon foods; *Cereus forbesii*; composition; native fruit; nutritional value; prickly pear

Citation: López, J.; Villalba, R.; Caballero, S.; Coronel, E.; Piris, P.; Friesen, A.; Mereles, L. Chaco Prickly Pear (*Cereus forbesii* Otto ex C.F. Först): An Ancient Source of Antioxidants and Dietary Fiber in the Diet of Indigenous Populations and Its Potential Application as an Ingredient in Derived Products. *Biol. Life Sci. Forum* **2022**, *17*, 21. https://doi.org/10.3390/blsf2022017021

Academic Editors: Norma Sammán, Mabel Cristina Tomás, Loreto Muñoz and Claudia Monika Haros

Published: 10 November 2022

1. Introduction

Despite the potential use of wild food plants in food security and poverty reduction strategies, the Food and Agriculture Organization of the United Nations (FAO) World Biodiversity Status for Food and Agriculture warns that this diversity is rapidly disappearing, unless the conservation of useful plants takes place. Only 3.3% of useful plants are sufficiently conserved *ex situ* [1]. Wild plant species play an important role in local and traditional food systems in rural communities. Research on such foodstuffs may help to prevent loss of indigenous knowledge on potential dietary sources for needy households [2]. The development of value-added products to incorporate them into the diet in a modern context can be a strategy for their conservation and incorporation into diet as high-value products or ingredients. For example, microencapsulation on alginate hydrogel has been reported to provide greater stability to betacyanin of dragon fruit peel extract, which allows the extraction and application of beneficial compounds as healthy food ingredients [3]. The CELAC Food Security Plan is based on four pillars that seek to guarantee the four dimensions of food security: access, availability, use and stability of food [4]. Indigenous people have a large amount of knowledge of their ecosystems, including knowledge of native foods, knowing its value and function in the maintenance of health and the control of different diseases with potential benefits for the welfare and health in the soil of the indigenous peoples, as well as for the industrialized populations [5]. The species *Cereus forbesii* is a perennial native tree distributed throughout Catamarca, Chaco, Córdoba, Formosa, Jujuy, La Rioja, Salta, Santa Fe, belongs to the Cactaceae family, considered synonymous with the species *Cereus validus* autc. Non Haw., *Piptanthocereus forbesii* (Otto ex C.F. Forst.) Riccob., nom. Illeg. [6] The pandemic opens up opportunities for a new paradigm of a food system that supports local self-sufficiency and domestic agricultural production and considers family and community gardens, traditional agroecosystems and farmers' markets as essential services. This work aims to describe the proximate composition and antioxidant potential of a wild prickly pear from Paraguayan Chaco and a derived product from them, marmalade.

2. Materials and Methods

2.1. Sampling

A sampling was carried out for the convenience of approximately 1000 kilos of tuna berries, harvested in Campo Loro, Misión Yalve Sanga, Kleefeld and Loma Plata, Philadelphia, Boquerón Department, Chaco, Paraguay at three harvest times, between December 2020 and January 2021. A random selection of fruits was carried out in good condition, approximately 1 kg per harvest. The samples were analyzed separately. From the same, 3 batches of marmalades made from whole fruits were prepared. The marmalade contained sugar and water addition.

2.2. Processing of the Samples

Once in the laboratory, the shell was separated from the pulp with seeds. These samples were freeze-dried for 10 days to eliminate the watery proportion, then they were ground in a food processor and kept frozen at −20 degrees centigrade until the moment of analysis. The morphological measurements of weight and size were carried out in the fresh sample. The analysis of the marmalades was carried out on fresh weight.

2.3. Analytical Methods

For morphological analysis, 30 whole fruits were taken, weight was measured in analytical balance (DNA, model HR 120), longitudinal and transversal diameter (measured in cm). For centesimal composition and vitamin C analysis, official AOAC techniques were used [7], dietary fiber AOAC 991.43 on pulp + seeds and AOAC 993.21 for marmalade, ash AOAC 940.26 method, total protein AOAC 920.152 method, carbohydrates by the Anthrone method [8] and the caloric value by the Atwater method. The total phenolic compounds (TCP) was carried out by the Folin–Ciocalteau method. The total antioxidant capacity (TAC) was carried out by the ABTS method described by Re et al. [9]. All reagents used were analytical-grade. All determinations were made in triplicate.

3. Results and Discussion

Physicochemical Characteristics and Antioxidants

The analyzed whole fruits of *Cereus forbesii* presented the following measurements: longitudinal diameter 9.44 ± 0.74 cm, transversal diameter 4.81 ± 0.35 cm and 104.7 ± 16.6 g of average weight (Figure 1). Table 1 summarizes the results of the analyzed samples. In its proximate composition, it highlights the high content of dietary fiber in pulp + seeds, as well as a great content of polyphenolic compounds with known antioxidant properties. In the marmalades, the composition of carbohydrates was the majority and the content of TPC was reduced by more than 80%. These results demonstrate the potential of the tuna del Chaco in the diets of indigenous communities, especially as fresh fruit in feeding programs. The incorporation in foods prepared at least as marmalades would be but a strategy to promote their consumption and as conservation systems and as a source of essential nutrients. The caloric contribution of wild food plants to the diets of people is generally low in comparison with basic foods. In the pulp samples with seeds, the caloric value was less than 70 Kcal/g. However, these species contribute to the diversification of the diet in many geographic environments. Additionally, the local trade in wild species has the potential to empower communities and increase living standards in rural areas within the framework of food security and sustainability of food systems.

(**a**)

(**b**)

Figure 1. Fruits of *Cereus forbesii*. (**a**). Whole fruit and marmalade (**b**). Cross section of a whole fruit.

Table 1. Physicochemical characteristics of fruits.

	***Cereus forbesii* Pulp + Seeds**	***Cereus forbesii* PPeel**	***Cereus forbesii* PMarmalade**
Moisture (g/100 g)	85.9–90.4	93.1–93.4	38.2–38.6
Carbohydrates (g/100 g)	2.5–3.72	2.45–2.59	50.7–56.8
Total Lipids (g/100 g)	tz	tz	tz
Proteins (g/100 g)	1.41–1.54	0.67–0.74	0.368–0.376
Ash (g/100 g)	0.09–0.17	0.66–0.71	0.64–0.92
Dietary fiber (g/100 g)	4.07–4.09	1.13–1.45	0.53–1.07
Caloric value (Kcal/100 g)	62–70	23–27	205–228
Vitamin C (mg/100 g)	1.22–1.23	0.49–0.93	tz
TPC (mg GAE/100 g)	334.7–387.6	349.4–509.8	46.41–60.91
ABTS (mM TEAC/g)	3.52–5.15	7.70–7.76	4.35–4.89

Results are expressed as intervals of n = 3 and triplicates. Where TPC = Total Phenolics Compounds, and Tz = trace.

4. Conclusions

The pulp and seeds of tuna represent a source of dietary fiber and polyphenolic compounds. The analyzed foods, especially pulp and its derivatives, have attractive color characteristics and antioxidant potential. Research on their micronutrients and phytochemical compounds can be interesting for the replacement of critical ingredients as artificial coloring additives in the diet of the regional population.

Author Contributions: Conceptualization, L.M. and J.L.; methodology, L.M.; software, E.C.; validation, R.V., P.P. and S.C.; formal analysis, J.L.; investigation, L.M. and A.F.; resources, L.M.; data curation, E.C.; writing—original draft preparation, J.L.; writing—review and editing, L.M.; visualization, R.V.; supervision, S.C.; project administration, L.M.; funding acquisition, L.M. All authors have read and agreed to the published version of the manuscript.

Funding: This work was supported by grant "Project For Our Great Sustainable Chaco: Active participation in territorial management models for environmental conservation integrated with sustainable production," executed by COOPI-Cooperazione Internazionale in consortium with CERDET and FCQ-

UNA, and the European Union "Conservation, sustainable use and good governance of biodiversity in four vulnerable biomes in central South America" (EuropeAid/154653/DD/ ACT/Multi).

Institutional Review Board Statement: Not applicable.

Informed Consent Statement: Not applicable.

Acknowledgments: The authors wish to thank Ia ValSe-Food-CYTED (119RT0567. Facultad de Ciencias Químicas and Tucos Factory E.I.R.L.

Conflicts of Interest: The authors declare no conflict of interest.

References

1. FAO—2014—El Estado de Los Recursos Genéticos Forestales En.Pdf. Available online: https://www.fao.org/3/i3827s/i3827s.pdf (accessed on 9 November 2022).
2. Rampedi, I.T.; Olivier, J. Traditional Beverages Derived from Wild Food Plant Species in the Vhembe District, Limpopo Province in South Africa. *Ecol. Food Nutr.* **2013**, *52*, 203–222. [CrossRef] [PubMed]
3. Bhagya Raj, G.V.S.; Dash, K.K. Microencapsulation of Betacyanin from Dragon Fruit Peel by Complex Coacervation: Physicochemical Characteristics, Thermal Stability, and Release Profile of Microcapsules. *Food Biosci.* **2022**, *49*, 101882. [CrossRef]
4. CEPAL-FAO-ALADI Seguridad Alimentaria, nutrición y erradicación del hambre. CELAC 2025. Elementos para el debate y la Cooperación regionales; Santiago, Chile. 2016. Available online: https://repositorio.cepal.org/bitstream/handle/11362/40348/1/S1600707_es.pdf (accessed on 9 November 2022).
5. Kuhnlein, H.; Erasmus, B.; Creed-Kanashiro, H.; Englberger, L.; Okeke, C.; Turner, N.; Allen, L.; Bhattacharjee, L. Indigenous Peoples' Food Systems for Health: Finding Interventions That Work. *Public Health Nutr.* **2006**, *9*, 1013–1019. [CrossRef] [PubMed]
6. Instituto de botánica Darwinion Catálogo de Plantas Vasculares Del Cono Sur (Argentina, Sur de Brasil, Chile, Paraguay y Uru guay). Available online: http://www.darwin.edu.ar/Proyectos/FloraArgentina/sinonimoespecie.asp?forma=&variedad=&subespecie=&EspCod=7581&especie=validus&genero=Cereus&SinGeneroDe=Cereus&SinEspecieDe=forbesii&sincod=7581 (accessed on 9 November 2022).
7. Horwitz, W.; Chichilo, P.; Reynols, H. *Official Methods of Analysis of the Association of Official Analytical Chemists*, 17th ed.; AOAC: Gaithersburg, MA, USA, 2000.
8. Dreywood, R. Qualitative Test for Carbohydrate Material. *Ind. Eng. Chem. Anal. Ed.* **1946**, *18*, 499. [CrossRef]
9. Re, R.; Pellegrini, N.; Proteggente, A.; Pannala, A.; Yang, M.; Rice-Evans, C. Antioxidant Activity Applying an Improved ABTS Radical Cation Decolorization Assay. *Free Radic. Biol. Med.* **1999**, *26*. [CrossRef]

Proceeding Paper

Obtaining Integral Kurugua Flour with Antioxidant Potential as Ingredient Foodstuffs †

Laura Correa [1], Marcia Gamarra [1], Kattya Sanabria [1], Eva Coronel [1], Marcela L. Martínez [2,3], Edgardo Calandri [2,4], Silvia Caballero [1] and Laura Mereles [1,*]

1 Facultad de Ciencias Químicas, Universidad Nacional de Asunción, San Lorenzo P.O. Box 1055, Paraguay
2 Instituto de Ciencia y Tecnología de los Alimentos (ICTA), Facultad de Ciencias Exactas, Físicas y Naturales (FCEFyN), Universidad Nacional de Córdoba (UNC), Córdoba 5000, Argentina
3 Instituto Multidisciplinario de Biología Vegetal (IMBIV), Consejo Nacional de Investigaciones Científicas y Técnicas (CONICET), Universidad Nacional de Córdoba (UNC), Córdoba 5000, Argentina
4 Instituto de Ciencia y Tecnología de Los Alimentos Córdoba (ICYTAC), Universidad Nacional de Córdoba (UNC), Córdoba 5000, Argentina
* Correspondence: lauramereles@qui.una.py
† Presented at the IV Conference Ia ValSe-Food CYTED and VII Symposium Chia-Link, La Plata and Jujuy, Argentina, 14–18 November 2022.

Abstract: The aim of the present work was to study the effect of the drying process conditions on the antioxidant properties of the integral fruit of kurugua (*Sicana odorifera* Naud.). The experiments showed that the antioxidant activity of fresh samples of whole kurugua could vary significantly depending on the fruit batch used. The statistical analyses showed no significant differences in antioxidant activity among the drying conditions studied. However, it is important to highlight that the drying process conducted at 80 °C and at an average air speed of 5.8 m/s presented the lowest cost (2.2 USD kW/h), and after 10 h, the raw material reached an aw level of 0.297, which is enough to inhibit the growth of microorganisms. As it is known, a low aw allows for a longer shelf-life of a product and prevents the proliferation of molds and yeasts. There was no significant difference in the concentration of β-carotene between drying times; nevertheless, the resulting flour showed a decrease in luminosity and color variation (b*) with respect to the fresh samples, with a typical browning due to the effects of temperature and air drying. The influence of the drying conditions on the integral kurugua flour is discussed in order to obtain the best dry product. A field of work has opened for future research on the sensory profile and its potential applications.

Keywords: autochthon foods; *Sicana odorifera*; composition; native fruit; nutritional value; flour

Citation: Correa, L.; Gamarra, M.; Sanabria, K.; Coronel, E.; Martínez, M.L.; Calandri, E.; Caballero, S.; Mereles, L. Obtaining Integral Kurugua Flour with Antioxidant Potential as Ingredient Foodstuffs. *Biol. Life Sci. Forum* **2022**, *17*, 22. https://doi.org/10.3390/blsf2022017022

Academic Editors: Norma Sammán, Mabel Cristina Tomás, Loreto Muñoz and Claudia Monika Haros

Published: 10 November 2022

1. Introduction

The *Sicana odorifera* Naud. fruit is considered a source of antioxidants and minerals [1]. However, despite it being a native fruit of Paraguay, its annual crops are underused due to low demand from consumers. The ignorance of its healthy properties limits its application in the diets of regional populations [2]. The kurugua crop was restored less than two decades ago as part of a biodiversity conservation strategy in Paraguay, promoting it as an alternative crop and reintroducing it into the population's food habits. Regionally, the pulps of black and red varieties of kurugua have been studied in relation to proteins, lipids, carbohydrates, and ashes. For the atropurpurea variety, the antioxidant properties of the cascara and its potential use as a source of natural dyes, in particular anthocyanins, have been studied. However, the development of products produced from the whole fruit is scarce, as is the study of its nutritional potential [3]. The aim of this work was to study the effect of the drying process conditions on the antioxidant properties of the integral fruit of kurugua (*Sicana odorifera* Naud.) in order to establish the best drying variable combinations to obtain kurugua flour. The results obtained enable the formulation of foods with a high nutritional quality, produced from the integral kurugua flour, with antioxidant potential.

2. Materials and Methods

2.1. Sampling

The fruits of *Sicana odorifera* were collected in a mature state and without visible damages in the 2021 harvest from a traditional crop by the Cordillera Department, Juan de Mena city (24°57′35.8″ S, 56°44′20.0″ W) Paraguay. Each sample had 20 kg of fruit.

2.2. Processing of the Samples

Once in the laboratory, the fruits were washed and cut into 1 cm-thick slices without separating the seeds and shells. The diameter and thickness were measured with a ruler and weighed in an analytical balance at the AND mark of each sample. A random sampling of 9 kurugua fruits was carried out, with an average of 1600 g of fresh fruit per drying experiment.

2.3. Analytical Methods

For the tray oven assays, the dryer was prepared by adjusting the temperature using a digital controller and the air speed using a manual angle valve, which was measured using the digital anemometer HoldPeak® model HP-866B (Zhuhai, China), according to the condition of the required work. Once the operating parameters were stabilized, the trays were placed inside the equipment, moving them each time a sample was performed, turning the slices by 180°. The operative conditions were performed in duplicate and were as follows: 80 °C—5.3 m/s (Cond. 1); 80 °C—3.8 m/s (Cond. 2); 60 °C—5.3 m/s (Cond. 3); 60 °C—3.8 m/s (Cond. 4).

The drying was performed using a tray-drying oven with forced air by a centrifugal fan and a transversal air inlet. The temperature was controlled by dry and wet bulbs provided at the dryer inlet and outlet. The lineal air speed, the mass flow rate and the relative humidity of the ambient air were controlled.

For the study of the drying conditions and kinetics, a longitudinal experimental design of 2^2 was used. The samples were taken every 60 min in order to determine the humidity, the aw, the total carotenoid content and the color.

For the chemical analysis, the wet content was measured at 45-min intervals for the samples taken from the dryer, and the humidity was analyzed on a RADWAG thermobalance model PM 60.3Y.WH (Torunska, Poland). A total antioxidant capacity (TAC) by an ABTS+ radical cation bleaching assay was carried out on the whole fruit before drying and on the dried product at its equilibrium moisture, in triplicates. For quantitation purposes, a calibration curve was provided with Trolox (aqueous solution 0–500 µM) at 730 nm [4]. The test was carried out spectrophotometrically at 730 nm. The water activity (aw), β-carotene and color were carried out every 2 h using the AOAC 978.19 method on a Rotronic HygroPalm equipment (New York, NY, USA). For the total carotenoid concentration, a spectrophotometric method at 450 nm [5] was employed, using the extraction solution as a blank, and, finally, calculating the concentration of carotene (c) in µg/mL. The color was determined by a standardized method using a ColorStay Colormeter software from Wuite Marten GMBH, 2020 (Baden-Wuttermberg, Germany) at an angle of 45° to the observer. The CIELAB parameters (L*, a* and b*) were selected to inform the color of the samples. All measurements were carried out in triplicate.

The energy cost estimation was calculated from the total drying time of each operational condition, as well as from the current intensities involved in the electrical resistance and the dryer fan. The voltage of the electrical installations was constant and equal to 220 V.

2.4. Statistical Analysis

The data were recorded and processed using the GraphPad Prism 8.2 program (GraphPad Software Inc., San Diego, CA, USA). To determine significant differences, a p value of ≤ 0.05 was considered.

3. Results and Discussion

3.1. Drying Curves

The time required for the fruit slices to reach equilibrium varied depending on the operating conditions. The results obtained from the drying curve under the different operating conditions show that the higher the temperature and air velocity, the lower the time to reach equilibrium humidity (Figure 1).

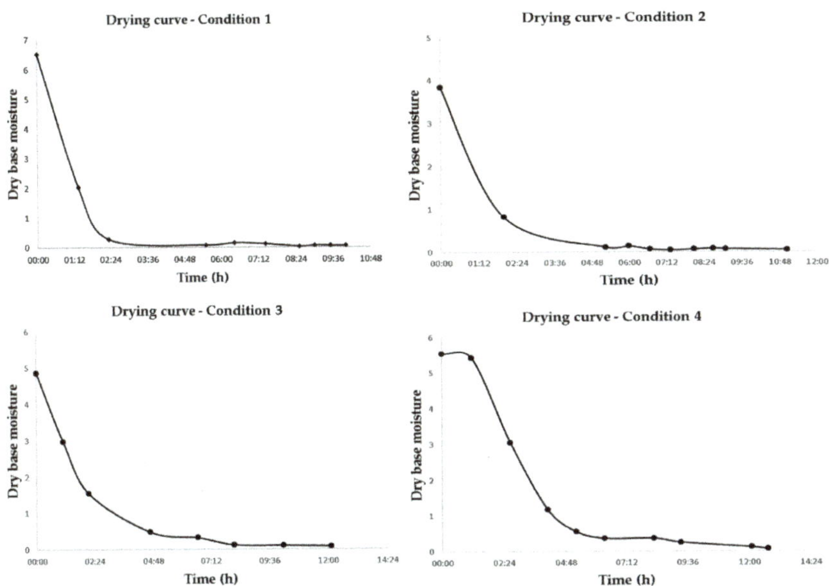

Figure 1. Kurugua fruit drying curves on moisture versus time for four conditions.

3.2. Influence of Temperature and Air Velocity on Total Antioxidant Capacity (TAC)

No significant differences were observed between the TAC values at the different drying conditions. This seems to indicate that the drying speed and the temperature do not affect the TAC of the final product in the evaluated conditions (Figure 2).

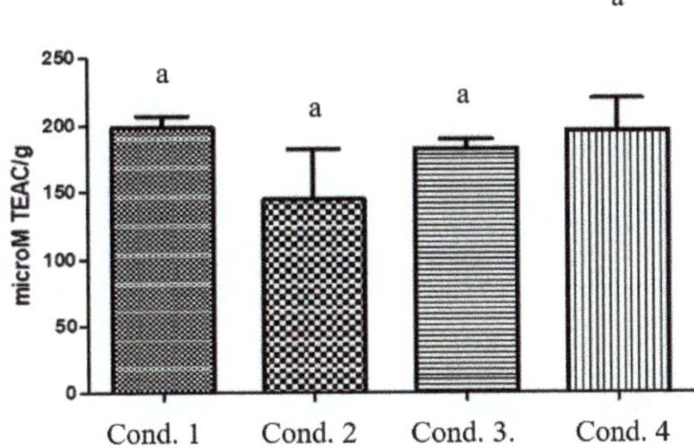

Figure 2. TAC corresponding to kurugua dried fruit flour under different operating conditions. The bars represent the means ± DS (n = 3). The different letters in the bars indicate significant differences between the measurements (single factor ANOVA, Tuckey post-test and $p \leq 0.05$), where Cond. 1 = 80 °C, 5.3 m/s; Cond. 2 = 80 °C, 3.8 m/s; Cond. 3 = 60 °C, 5.3 m/s; Cond. 4 = 60 °C, 3.8 m/s.

The drying times to reach the humidity equilibrium were 9.52 h, 11.2 h, 12.08 h and 12.43 h under Cond. 1 (80 °C, 5.3 m/s), Cond. 2 (80 °C, 3.8 m/s), Cond. 3 (60 °C, 5.3 m/s) and Cond. 4 (60 °C, 3.8 m/s), respectively. As expected, Cond. 1 presented the lowest average drying time and energy cost (2.2 USD kW/h). The aw values showed that 10 h is enough to reach a level of aw (0.297) that inhibits the growth of microorganisms. It is well known that low aw values allow for a longer shelf-life of a product and inhibit the proliferation of molds and yeast [6]. There were no significant differences in the total carotenoid concentrations in the 0–10-h drying range (Figure 3). However, the resulting flour showed a decrease in luminosity and color variation (b*) with respect to the fresh samples, with a typical browning due to the effect of the temperature and air during the drying. In summary, the drying operation conditions to obtain a dry product from kurugua with an acceptable appearance and antioxidant capacity are described. The basis for obtaining a dry product, integral kurugua flour, from experimental drying conditions has been established, and a field of work has been opened for future research on the sensory profile and its potential applications.

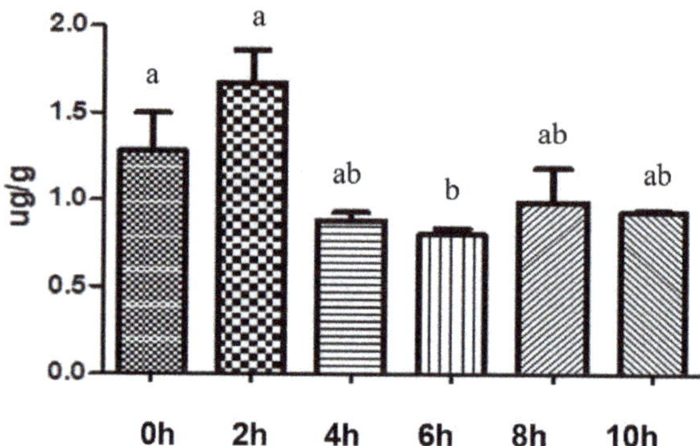

Figure 3. Total carotenoid concentrations at different drying times. The bars represent the means ± DS (n = 3). The different letters in the bars indicate significant differences between the measurements (single factor ANOVA, Tukey test for different comparisons and $p \leq 0.05$).

4. Conclusions

There were no significant variations in the total antioxidant capacity and total carotenoid content between the four drying conditions used in this study. A temperature of 80 °C and an average air speed of 5.8 m/s were chosen due to the low operating cost and adequate aw. Based on these results, future studies can be carried out on the application of this product as a natural ingredient.

Author Contributions: Conceptualization, L.M. and E.C. (Edgardo Calandri); methodology, M.L.M. and L.C.; software, E.C. (Eva Coronel); validation, L.C., M.G. and K.S.; formal analysis, K.S.; investigation, L.C.; resources, S.C.; data curation, L.M.; writing—original draft preparation, L.C.; writing—review and editing, L.M.; visualization, M.L.M.; supervision, L.M., M.L.M. and E.C. (Edgardo Calandri); project administration, S.C.; All authors have read and agreed to the published version of the manuscript.

Funding: This work was supported by the grant Ia ValSe-Food-CYTED (119RT0567), and Blas Imas from the "Kurugua Poty" Foundation.

Institutional Review Board Statement: Not applicable.

Informed Consent Statement: Not applicable.

Data Availability Statement: Not applicable.

Conflicts of Interest: The authors declare no conflict of interest.

References

1. Albuquerque, B.R.; Dias, M.I.; Pereira, C.; Petrović, J.; Soković, M.; Calhelha, R.C.; Oliveira, M.B.P.P.; Ferreira, I.C.F.R.; Barros, L. Valorization of Sicanaodorifera (Vell.) Naudin Epicarp as a Source of Bioactive Compounds: Chemical Characterization and Evaluation of Its Bioactive Properties. *Foods* **2021**, *10*, 700. [CrossRef] [PubMed]
2. Caballero, S.; Mereles, L.; Burgos-Edwards, A.; Alvarenga, N.; Coronel, E.; Villalba, R.; Heinichen, O. Nutritional and Bioactive Characterization of Sicana Odorifera Naudim Vell. Seeds By-Products and Its Potential Hepatoprotective Properties in Swiss Albino Mice. *Biology* **2021**, *10*, 1351. [CrossRef] [PubMed]
3. Mereles, L.; Caballero, S.; Burgos-Edwards, A.; Benítez, M.; Ferreira, D.; Coronel, E.; Ferreiro, O. Extraction of Total Anthocyanins from Sicana Odorifera Black Peel Fruits Growing in Paraguay for Food Applications. *Appl. Sci.* **2021**, *11*, 6026. [CrossRef]
4. Re, R.; Pellegrini, N.; Proteggente, A.; Pannala, A.; Yang, M.; Rice-Evans, C. Antioxidant Activity Applying an Improved ABTS Radical Cation Decolorization Assay. *Free Radic. Biol. Med.* **1999**, *26*, 1231–1237. [CrossRef]
5. Murillo, E.; Meléndez-Martínez, A.J.; Portugal, F. Screening of Vegetables and Fruits from Panama for Rich Sources of Lutein and Zeaxanthin. *Food Chem.* **2010**, *122*, 167–172. [CrossRef]
6. do Nascimento Nunes MC Impact of Environmental Conditions on Fruit and Vegetable Quality. *Stewart Postharvest Rev.* **2008**, *4*, 1–14. [CrossRef]

Proceeding Paper

Characterization of Three-Layer Microcapsules of Chia Seed Oil Obtained for Electrostatic Deposition Technology †

Claudia N. Copado [1], Luciana M. Julio [1], Bernd W. Diehl [2], Vanesa Y. Ixtaina [1,3] and Mabel C. Tomás [1,*]

1 Centro de Investigación y Desarrollo en Criotecnología de Alimentos (CIDCA), CCT La Plata (CONICET), Facultad de Ciencias Exactas (FCE), Comisión de Investigaciones Científicas de la Provincia de Buenos Aires (CICPBA), Universidad Nacional de La Plata (UNLP), calle 47 y 116, La Plata CP1900, Argentina
2 Spectral Service GmbH, Emil Hoffman Str. 33, 50996 Cologne, Germany
3 Facultad de Ciencias Agrarias y Forestales (FCAyF), Universidad Nacional de La Plata (UNLP), calle 60 y 119, La Plata CP1900, Argentina
* Correspondence: mabtom@hotmail.com
† Presented at the IV Conference Ia ValSe-Food CYTED and VII Symposium Chia-Link, La Plata and Jujuy, Argentina, 14–18 November 2022.

Abstract: Oils with high omega-3 fatty acids are known for their multiple health benefits. For this reason, new strategies have been developed to protect these fatty acids from lipid oxidation to incorporate them into foods. Microencapsulation is an attractive alternative for protecting and incorporating chia oil in food matrices. A three-layer microencapsulation process was performed using a layer-by-layer technique with the addition of sunflower lecithin, chitosan, and chia mucilage by spray-drying. The microcapsules obtained were studied and stored in darkness at a controlled temperature and relative humidity for 90 days. The ζ-potential evidenced the electrostatic deposition of the layers in powders through the change of the values in the electric charge. Microcapsules showed a high microencapsulation efficiency and low moisture content and water activity levels. Microencapsulated chia oil presented low oxidation values (<10 meq hydroperoxides/kg oil) and high omega-3 fatty acid content after storage. These results suggest that the three-layer microcapsules studied are suitable to provide high stability against the oxidative deterioration of functional lipid components in chia oil and constitute a promising application in the food industry.

Keywords: chia oil; α-linolenic acid; LBL technique; microcapsules; Omega-3

Citation: Copado, C.N.; Julio, L.M.; Diehl, B.W.; Ixtaina, V.Y.; Tomás, M.C. Characterization of Three-Layer Microcapsules of Chia Seed Oil Obtained for Electrostatic Deposition Technology. *Biol. Life Sci. Forum* **2022**, *17*, 23. https://doi.org/10.3390/blsf2022017023

Academic Editors: Norma Sammán, Loreto Muñoz and Claudia Monika Haros

Published: 10 November 2022

1. Introduction

Currently, oils rich in ω-3 fatty acids are known for their multiple health benefits. For this reason, new functional ingredients are being developed that allow their incorporation into foods and protect the oil from lipid oxidation. Microencapsulation represents an attractive alternative for protecting and incorporating chia oil in various food matrices [1]. One of the most commonly used encapsulation methods is spray-drying [2]. Obtaining stable emulsions with a subsequent proper drying process for conversion into powder material is essential in preparing microencapsulated oils. Different factors, such as the characteristics of the parent oil-in-water (O/W) emulsion, the total solid content, the type of wall and core material, and the microencapsulation processing conditions, can affect the physicochemical properties and the stability of the oil microparticles [2]. The layer-by-layer (LBL) technique is a strategy to improve the emulsion stability, which enables the creation of multiple layers of emulsifiers and polyelectrolytes around the oil droplets [1]. Sunflower lecithin can be used as an emulsifier and presents the advantage that it is a non-GMO additive, which some consumers appreciate. Chitosan soluble in acidic aqueous media presents a positive charge at pH values < 8, increasing its ζ-potential as the pH decreases. Chia mucilage, an anionic polymer within a pH range of 1.8–12.0, confers viscosity to dispersions in water [3].

The objective of this work was to obtain chia oil three-layer microcapsules by spray-drying using modified sunflower lecithins, chitosan, chia mucilage, and maltodextrin as wall materials, carrying out the respective physicochemical characterization.

2. Materials and Methods

2.1. Materials

Cold-pressed chia oil was provided by Solazteca SDA (Lobos, Buenos Aires, Argentina). The modified (deoiled (LD) and hydrolyzed (LH)) sunflower lecithins were provided by Lasenor Emul S.L. (Olesa de Montserrat, Barcelona, Spain). Chitosan (Q) was purchased from Sigma Chemical Company (St. Louis, MO, USA). Maltodextrin (Mxd) DE 13–17% was donated by INDECAR SAICyF (Ciudad Autónoma de Buenos Aires, Argentina). Chia mucilage was extracted from the seed, as described by Copado et al., 2021 [1].

2.2. Methods

2.2.1. Parent Emulsions: Preparation and ζ-Potential Determination

The stock dispersions to prepare the emulsions were: LD or LH 2.2% w/w, Q 2.0% w/w, and chia mucilage 11.9% w/w (pH 5). Pre-emulsification was made using an Ultraturrax homogenizer (IKA-Labortechnik, 130 GmbH & Co., Staufen, Germany), 2 min at 9500 rpm. Subsequently, a high-pressure valve homogenizer (Panda 2K GEA Niro Soavi, Parma, Italy) (1000 bar, 4 cycles) was used to obtain the primary. The chitosan dispersion addition using similar operating conditions allowed for obtaining the secondary emulsion. After that, chia mucilage was added through an ultrasonic processor (VCX 750, Sonics & Materials Inc. Newtown, CT, USA) (amplitude of 40% for 2 min). Finally, Mxd was incorporated. The total solid content was 26% w/w (final concentration in emulsion: 5% chia oil, 0.5% lecithins (LD or LH), 18.2% Mx, 0.3% chitosan, and 2% chia mucilage).

The ζ-potential after each step of the three-layer emulsion obtention was analyzed using a zeta potential analyzer (Brookhaven 90Plus/Bi-MAS, Holtsville, NY, USA) on the function of electrophoretic mobility at room temperature [4].

2.2.2. Microcapsule Preparation by Spray-Drying

The spray-drying of the emulsions was carried out in a Mini Spray Dryer B-290 (BÜCHI, Flawil, Switzerland) (0.5 mm diameter nozzle, 0.6 L/h of feed rate, and 170/75 °C of air inlet/outlet temperatures) [5].

The storage of the microcapsules was made at 33%RH, 25 ± 2 °C, in darkness for 90 d. Bulk chia oil stored under the same conditions was included as control system.

2.2.3. Microcapsule Characterization

Moisture Content and Water Activity (a_w)

The moisture content and the a_w was evaluated, according to Copado et al. [1].

Microencapsulation Efficiency

It was calculated by the relation between the encapsulated and the total oil, both gravimetrically determined, according to Copado et al. [1] and Klinkesorn et al. [6], respectively.

ζ-potential of the Reconstituted Emulsions

The powders were reconstituted in a ratio of 10% w/w with acetic acid/acetate buffer (100 mM) at a pH of 5, at ~25 °C, stirring for 30 min. Afterward, the ζ-potential was analyzed, according to the procedure detailed previously for parent emulsions.

Peroxide Value (PV)

The primary lipid oxidation products were evaluated, according to the method of Díaz et.al. (2003) [7].

Content of Omega-3 Polyunsaturated Fatty Acids (PUFAs)

It was analyzed by NMR spectroscopy using an Avance III 400 MHz spectrometer (Bruker Biospin, Rheinstetten, Germany) with a 5 mm BBI probe (Eurofins, Hamburg, Germany) [8]. Spectroscopic measurements (^{1}H NMR) were carried out in triplicate.

2.2.4. Statistical Analysis

To detect any significant difference between samples, data were subjected to analysis of variance (ANOVA) at a 95% confidence level ($p \leq 0.05$). Significantly different data sets were classified after post hoc comparison tests using Tukey´s honestly significant differences test (HSD, $p \leq 0.05$).

3. Results and Discussion

3.1. *ζ-Potential of the Parent Emulsions*

The deposition of layers was verified by the inversion of the electrical charge after each step of the parent emulsion preparation [9]. After the lecithin addition, the emulsions presented a negative ζ-potential (−10.39 and −6.43 mV for emulsion with LD and LH, respectively). Positive values of ζ-potential were recorded after the addition of Q (31.08 mV for LD, and 28.99 for LH). This fact shows the successful chitosan deposition on the layer of modified sunflower lecithins at a pH of 5, conferring to the particles a cationic character. Finally, a new change of ζ-potential towards negative values was recorded for the three-layer emulsions, caused by the interaction between the chitosan and the chia mucilage forming the third layer (−18.91 mV for LD; −18.87 mV for LH). These results are according to other research works, which reported changes in the electrical charges during the formulation of multilayered microparticles by the LBL electrostatic deposition technique [10,11].

3.2. *Characterization of Multilayer Microcapsules*

3.2.1. Moisture Content (MC%) and Water Activity (a_w)

The values of MC% and a_w (Table 1) varied between 0.86–1.29% and 0.224–0.280, respectively, which are suitable values for powder products (3–4% d.b.) [1]. No significant differences ($p > 0.05$) were recorded in MC% for the microcapsules formulated with both lecithin types, whereas the a_w of TLD was significantly ($p \leq 0.05$) higher than TLH. The a_w values are in the range where the lipid oxidation is minimal (0.2–0.4) due to a delay in the decomposition of the hydroperoxides and a decrease in the pro-oxidant activity of metals that promotes the oxidation process [12].

Table 1. Physicochemical characterization of chia oil microcapsules at t = 0 and 90 d of storage at 25 ± 2 °C, 33% RH, darkness.

Microcapsule	MC% (d.b.)	a_w (25 °C)	ME	PUFAs Omega-3 Content	
				T = 0 d	T = 90 d
TLD	0.86 ± 0.16 ^{a}	0.280 ± 0.003 ^{b}	98.33 ± 0.51 ^{a}	60.40 ± 0.00 ^{aA}	60.35 ± 0.64 ^{aA}
TLH	1.29 ± 0.04 ^{a}	0.224 ± 0.015 ^{a}	98.16 ± 0.52 ^{a}	60.50 ± 0.00 ^{aA}	60.70 ± 0.14 ^{aA}

Mean values ± standard deviation (n = 3). The different letters in each column indicate differences ($p \leq 0.05$) between systems, according to the Tukey test (HSD). Different capital letters in the rows indicate differences between storage times ($p \leq 0.05$) for each system, according to the Tukey test (HSD). Moisture content (MC%), dry basis (d.b.), water activity (a_w), microencapsulation efficiency (ME), and polyunsaturated fatty acid content (PUFAS omega-3 content) at initial time (t = 0 d) and final time (t = 90 d). TLD: Three layer with LD, TLH: Three layer with LH.

3.2.2. Microencapsulation Efficiency (ME)

The studied systems presented high ME values (98.33–98.16%), indicating that the wall materials and the microencapsulation process were appropriate to encapsulate the lipid nucleus. No significant differences ($p > 0.05$) were found between the microcapsules with different modified lecithin. It could be related to the multiple layers of chitosan and mucilage that covered up any possible effect of the lecithin type.

3.2.3. ζ-Potential

The ζ-potential of the particles of the reconstituted emulsions was similar to the parent emulsions (TLD-15.45 mV and TLH-15.90 mV). These results suggest different layers in the powders showing that the microencapsulation process did not alter the electrical nature of the biopolymers.

3.2.4. Oxidative Stability: PV Values and PUFAs Omega-3 Content

At the initial time, the PV were 1.02, 1.06, and 2.02 meq of peroxide/kg of oil for TLD, TLH, and chia oil, respectively. The PV of the bulk chia oil increased faster than the microencapsulated systems during storage at 33% RH, 25 ± 2 °C, in darkness, achieving a value of 18.99 meq of peroxide/kg of oil after 90 d. On the contrary, the microencapsulated oil did not present significant differences ($p > 0.05$) on storage time, with PV < 10 meq/kg oil being the maximum value allowed for the consumption of chia oil (Codex Alimentarius Commission, Standard CXS 19-1981, amended in 2019).

Regarding the ω-3 PUFAs of the microcapsules, they did not show significant variation between the initial and the final storage time. The content of these PUFAs was ~60%, without significant differences ($p > 0.05$) between the systems (Table 1). These results are analogous to the PV, suggesting that the three-layer microcapsules are efficient systems to protect chia oil against lipid oxidation. The chia mucilage addition would increase the thickness of the interfacial membrane, constituting an additional barrier to oxygen diffusion.

4. Conclusions

In the present research work, chia oil microcapsules were developed from the spray-dying of three-layer emulsions obtained by electrostatic deposition, using the LBL technique. The results allowed us to observe a high microencapsulation efficiency, suggesting the process that the materials used to form the wall were adequate to encapsulate the oil. The spray-drying conditions were appropriate, since the moisture and a_w values are within the accepted limits for the stability of dehydrated foods. The electrical charge of the rehydrated systems allowed us to verify the different layers formed in the microcapsules. The peroxide values showed a marked increase in the oxidation of the unencapsulated oil, compared with those of the microencapsulated systems, which were below 10 meq of hydroperoxides/kg of oil. Therefore, the studied microcapsules were efficient protecting systems against the oxidation of chia oil-sensitive compounds. This protective effect can also be seen in the high omega-3 content of microcapsules at the end of storage.

These results suggest that the three-layer microcapsules are suitable for providing high oxidation stability of the functional lipid components in chia oil and are interesting for their application in the food industry for food enrichment.

Author Contributions: Conceptualization, Methodology, Investigation, Formal analysis, Writing—original draft preparation, Visualization C.N.C.; Conceptualization, Methodology, Investigation, Formal analysis L.M.J.; Methodology, Formal analysis B.W.D.; Conceptualization, Methodology, Investigation, Formal analysis, Writing—review & editing, Supervision, Resources V.Y.I.; Conceptualization, Methodology, Investigation, Formal analysis, Writing—review & editing, Supervision, Resources, Project administration, Funding acquisition M.C.T. All authors have read and agreed to the published version of the manuscript.

Funding: This research was funded by laValSe-Food-CYTED (119RT0567), Universidad Nacional de La Plata (UNLP) (X907), Agencia Nacional de Promoción Científica y Tecnológica (PICT 2016-0323, 2020-01274), Consejo Nacional de Investigaciones Científicas y Técnicas (CONICET) PIP 2007 (Argentina).

Institutional Review Board Statement: Not applicable.

Informed Consent Statement: Not applicable.

Data Availability Statement: Data available on request.

Conflicts of Interest: The authors declare no conflict of interest.

References

1. Copado, C.N.; Julio, L.M.; Diehl, B.W.; Ixtaina, V.Y.; Tomás, M.C. Multilayer microencapsulation of chia seed oil by spray-drying using electrostatic deposition technology. *LWT* **2021**, *152*, 112206. [CrossRef]
2. Augustin, M.A.; Sanguansri, L. Microencapsulation technologies. In *Engineering Foods for Bioactives Stability and Delivery*; Springer: New York, NY, USA, 2017; pp. 119–142.
3. Timilsena, Y.P.; Adhikari, R.; Barrow, C.J.; Adhikari, B. Physicochemical and functional properties of protein isolate produced from Australian chia seeds. *Food Chem.* **2016**, *212*, 648–656. [CrossRef] [PubMed]
4. Kwamman, Y.; Mahisanunt, B.; Matsukawa, S.; Klinkesorn, U. Evaluation of electrostatic interaction between lysolecithin and chitosan in two-layer tuna oil emulsions by nuclear magnetic resonance (NMR) spectroscopy. *Food Biophys.* **2016**, *11*, 165–175. [CrossRef]
5. Us-Medina, U.; Julio, L.M.; Segura-Campos, M.R.; Ixtaina, V.Y.; Tomás, M.C. Development and characterization of spray-dried chia oil microcapsules using by-products from chia as wall material. *Powder Technol.* **2018**, *334*, 1–8. [CrossRef]
6. Klinkesorn, U.; Sophanodora, P.; Chinachoti, P.; McClements, D.J.; Decker, E.A. Stability of spray-dried tuna oil emulsions encapsulated with two-layered interfacial membranes. *J. Agric. Food Chem.* **2005**, *53*, 8365–8371. [CrossRef] [PubMed]
7. Díaz, M.; Dunn, C.M.; McClements, D.J.; Decker, E.A. Use of caseinophosphopeptides as natural antioxidants in oil-in-water emulsions. *J. Agric. Food Chem.* **2003**, *51*, 2365–2370. [CrossRef] [PubMed]
8. Copado, C.N.; Diehl, B.W.; Ixtaina, V.Y.; Tomás, M.C. Application of Maillard reaction products on chia seed oil microcapsules with different core/wall ratios. *LWT* **2017**, *86*, 408–417. [CrossRef]
9. McClements, D.J.; Weiss, J. Lipid emulsions. In *Bailey's Industrial Oil and Fat Products*; John Wiley & Sons: Hoboken, NJ, USA, 2005.
10. Julio, L.; Copado, C.; Crespo, R.; Diehl, B.; Ixtaina, V.; Tomás, M.C. Design of microparticles of chia seed oil by using the electrostatic layer-by-layer deposition technique. *Powder Technol.* **2019**, *345*, 750–757. [CrossRef]
11. Noello, C.; Carvalho, A.; Silva, V.; Hubinger, M. Spray dried microparticles of chia oil using emulsion stabilized by whey protein concentrate and pectin by electrostatic deposition. *Food Res. Int.* **2016**, *89*, 549–557. [CrossRef] [PubMed]
12. Frankel, E. *Lipid Oxidation*, 2nd ed.; The Oily Press: Bridgwater, UK, 2005.

Proceeding Paper

Molecular Encapsulation of Hydrolyzed Chia Seed Oil by Ultrasonically Treated Amylose Inclusion Complexes †

Andrea E. Di Marco [1], Vanesa Y. Ixtaina [1,2] and Mabel C. Tomás [1,*]

1 Centro de Investigación y Desarrollo en Criotecnología de Alimentos (CIDCA), Consejo Nacional de Investigaciones Científicas y Técnicas (CONICET), Facultad de Ciencias Exactas (FCE), Comisión de Investigaciones Científicas de la Provincia de Buenos Aires (CICPBA), Universidad Nacional de La Plata (UNLP), calle 47 y 116, La Plata 1900, Argentina

2 Facultad de Ciencias Agrarias y Forestales (FCAyF), Universidad Nacional de La Plata (UNLP), calle 60 y 119, La Plata 1900, Argentina

* Correspondence: mabtom@hotmail.com

† Presented at the IV Conference Ia ValSe-Food CYTED and VII Symposium Chia-Link, La Plata and Jujuy, Argentina, 14–18 November 2022.

Abstract: Chia (*Salvia hispanica* L.) seed oil is a naturally rich source of α-linolenic (~65%) and linoleic (~20%) essential fatty acids, which are known for their beneficial effects on health. However, they are highly susceptible to oxidative deterioration. Amylose, the linear component of starch, has the ability to form inclusion complexes with hydrophobic molecules (ligand), which may act as delivery systems of sensitive bioactive compounds, including essential omega-3 and omega-6 fatty acids. In the present work, the hydrolytic effectiveness of *Candida rugosa* and porcine pancreatic lipases to obtain chia seed oil-free fatty acids was compared, which were complexed with high-amylose starch through the alkaline method with and without the incorporation of ultrasonic treatment. The highest level of free fatty acids released (>80%) was reached with *Candida rugosa* lipase. The inclusion complexes formed with this hydrolysate displayed a typical V-type X-ray diffraction pattern (peaks at ~7.5, 13, and 20° (2θ)), which confirmed an effective complexation. Moreover, ultrasonically treated complexes displayed a small peak at ~21°, from crystallized saturated fatty acids. Through attenuated total reflectance Fourier-transform infrared spectroscopy, the presence of typical bands of fatty acids in the complexes was verified, whose intensity increased after the application of ultrasonic treatment. The dissociation temperature determined using differential scanning calorimetry was >90 °C. According to this, *Candida rugosa* lipase showed better hydrolytic effectiveness on chia seed oil, and the fatty acids released were able to form amylose inclusion complexes with high thermal stability, whose properties varied after ultrasonic treatment.

Keywords: amylose inclusion complex; chia seed oil; enzymatic hydrolysis; α-linolenic acid; linoleic acid; ultrasound

Citation: Di Marco, A.E.; Ixtaina, V.Y.; Tomás, M.C. Molecular Encapsulation of Hydrolyzed Chia Seed Oil by Ultrasonically Treated Amylose Inclusion Complexes. *Biol. Life Sci. Forum* **2022**, *17*, 24. https://doi.org/10.3390/blsf2022017024

Academic Editors: Norma Sammán, Loreto Muñoz and Claudia Monika Haros

Published: 10 November 2022

1. Introduction

Chia (*Salvia hispanica* L.) seed oil is a rich source of α-linolenic (C18:3, ~60%) and linoleic (C18:2, ~18%) fatty acids (FAs). Their multiple health benefits and susceptibility to oxidative deterioration have promoted the development of delivery systems. Amylose, the linear component of starch, interacts with hydrophobic molecules forming inclusion complexes (ICs), a helical structure that incorporates the guest (ligand) inside its inner cavity, which may potentially act as carrier agents of bioactive compounds [1]. The complexation of triacylglycerols is restricted by steric hindrance effects and low water solubility [2], while the amylose interaction with free fatty acids may be enhanced through the application of ultrasonic treatment (UT) [3].

The aim of the present work was to compare the effectiveness of porcine pancreatic (PP) and *Candida rugosa* (CR) lipases to hydrolyze chia seed oil and to study the physicochemical properties of amylose–chia seed oil fatty acid ICs formed with and without the application of ultrasonic treatment. It will contribute to the development of ICs as potential delivery systems of chia oil essential fatty acids.

2. Materials and Methods

2.1. Materials

High-amylose corn starch (>70% amylose) was kindly provided by Ingredion Inc. (Westchester, IL, USA). Chia oil (fatty acid composition: α-linolenic 63.5%, linoleic 18.2%, palmitic 9.4%, oleic 5.7%, and stearic 3.2% acids) was purchased from Solazteca SDA S.A. (Buenos Aires, Argentina). Porcine pancreatic (type II, enzymatic activity 100–400 U/mg) and *Candida rugosa* (type VII, enzymatic activity ≥700 U/mg) lipases were purchased from Sigma-Aldrich (St. Louis, MO, USA). All other reagents used were of analytical grade.

2.2. Chia Seed Oil Enzymatic Hydrolysis

The hydrolytic reaction of chia oil with porcine pancreatic and *Candida rugosa* lipases was performed and monitored according to the procedures previously described [4], i.e., o/w emulsion 20% *w/w*, 37 °C, pH = 7, free fatty acid (FFA, %) measurement by titration with 0.1N NaOH. Then, the fatty acids released (hydrolysate) were extracted from the reaction mixture that reached the highest FFA content, by acidification to pH < 2 with 4N HCl, followed by centrifugation (4000× *g*, 20 min).

2.3. Formation of Untreated and Ultrasonically Treated Inclusion Complexes

The inclusion complex formation was performed following the procedures previously described [4], with a crystallization step of 2 h at 90 °C. Ultrasonically treated complexes were obtained under the same protocol, but subjecting the starch–hydrolysate mixture to ultrasound (1 min, 30% amplitude, pulsed on/off for 5 s) before their acidification to pH~4.7, using a VCX 750 ultrasonic processor (Sonics & Materials Inc., Newtown, CT, USA).

2.4. Attenuated Total Reflectance Fourier-Transform Infrared Spectroscopy (ATR-FTIR)

ATR-FTIR spectra of powdered ICs were directly measured and recorded in the wavenumber range of 500 to 4000 cm^{-1} under 4 cm^{-1} spectral resolution accumulating 16 scans per spectra, using an ATR-FTIR Thermo Nicolet iS10 spectrometer (Thermo Scientific, Waltham, MA, USA). The results were analyzed using OMNIC software (version 8.3, Thermo Scientific, MA, USA).

2.5. X-ray Diffraction (XRD)

The X-ray diffraction measurements of powdered ICs were carried out using a PANalytical X'Pert Pro diffractometer (Panalytical, Netherlands). Operating conditions: CuKα radiation (λ = 1.5403 Å), current 40 mA, voltage 40 kV, 5–35° (2θ) range, steps of 0.02°/s. The diffractograms were analyzed using the PeakFit v4.12 software (SeaSolve Software Inc., San José, CA, USA) and the crystallinity (%) was calculated as the ratio of the crystalline to the total (crystalline + amorphous) peak areas [4].

2.6. Differential Scanning Calorimetry (DSC)

The endothermic transitions of untreated and ultrasonically treated ICs previously suspended distilled water (1/4 *w/w* IC/water) were registered according to the procedure described by [4] (~8 mg of sample in hermetically sealed aluminum pans, 20–150 °C range at 5 °C/min) with a differential scanning calorimeter Q100 (TA Instruments, New Castle, DE, USA). A sealed empty pan was used as the reference. Thermograms were analyzed with the TA Instruments Universal Analysis 2000 software (TA, New Castle, DE, USA).

2.7. Statistical Analysis

ANOVA and Tukey ($p \leq 0.05$) tests were used to establish significant differences between means, and performed using the Statgraphics Centurion XV.II software (StatPoint Technologies, Warrenton, VA, USA).

3. Results and Discussion

3.1. Enzymatic Hydrolysis of Chia Seed Oil

Figure 1 shows the free fatty acids (%) released by *Candida rugosa* and porcine pancreatic lipases. As can be seen, in both cases, a rapid increase in FFA was observed at the beginning of the reaction, followed by a plateau in which the FFA remained almost constant over time. The maximum values reached by CR and PP lipases after 5 h of reaction were 82 and 24% FFA, respectively, indicating that CR had better hydrolytic effectiveness than PP lipase (Figure 1). The microbial enzyme catalyzes the hydrolysis of triacylglycerols randomly and is, therefore, able to release all types of acyl chains, regardless of their position in the glycerol molecule. The animal enzyme is a typical sn-1,3-specific lipase, i.e., the hydrolysis of triacylglycerols occurs in both the *sn*-1 and *sn*-3 positions of the glycerol backbone, while the fatty acid esterified in the *sn*-2 position remains non-hydrolyzed [5]. This may explain the better effectiveness of CR lipase than PP during the chia seed oil hydrolysis. Based on these results, the hydrolyzed lipid fraction obtained using the enzyme from CR was chosen to be used as a ligand during the following complex formation.

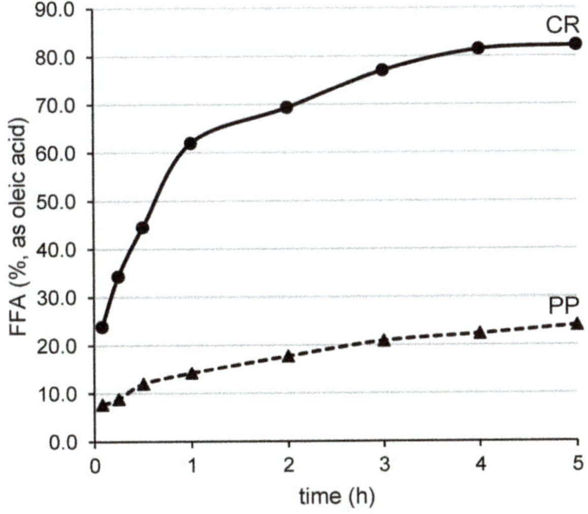

Figure 1. Free fatty acids (%, as oleic acid) released by enzymatic hydrolysis of chia seed oil with *Candida rugosa* (CR) and porcine pancreatic (PP) lipases.

3.2. Attenuated Total Reflectance Fourier-Transform Infrared Spectroscopy (ATR-FTIR)

The original (non-hydrolyzed) chia seed oil displayed a high-intensity band at ~1742 cm^{-h} (Figure 2a), from the stretching of the ester bonding between the glycerol and fatty acids in the triacylglycerol molecule. This band was not observed in the hydrolysate spectrum, but a new high-intensity band appeared at ~1707 cm^{-1} (Figure 2b), which originated from the stretching vibration of the carbonyl (–C = O) of the acid functional group of FAs released by the enzymatic hydrolysis. Moreover, the = C – H vibration of cis double bonds of unsaturated FAs originated a medium-intensity band at ~3010 cm^{-1} in both the oil and hydrolysate (Figure 2a,b). These samples also showed bands at 2852 and 2924 cm^{-1}, corresponding to the symmetric and asymmetric vibration of the CH_2 groups from the fatty acid alkyl chains, respectively (Figure 2a,b).

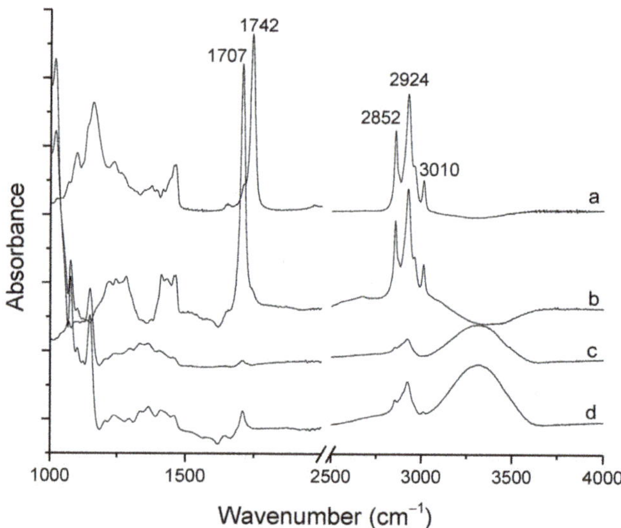

Figure 2. ATR-FTIR spectra of: (a) chia seed oil, (b) enzymatically hydrolyzed chia seed oil, (c) untreated inclusion complexes, and (d) ultrasonically treated inclusion complexes.

The typical bands from the hydrolysate previously mentioned were also present with lower intensity in the spectra of ICs, especially those at 1707, 2852, and 2924 cm^{-1} (Figure 2c,d). It confirms the presence of guest FAs in the ICs formed under the different conditions studied. The peak intensity in the ultrasonically treated ICs was higher than in the untreated ones, suggesting higher retention of FAs after ultrasonic treatment.

3.3. X-ray Diffraction

The crystalline structure of the powdered ICs obtained after freeze-drying was characterized by XRD. Both untreated and ultrasonically treated complexes displayed a semi-crystalline V-type diffraction pattern with two main reflections at ~13 and 20° (2θ) and a small reflection at ~7.5° (Figure 3), confirming an effective complexation of chia oil fatty acids. In addition, the ultrasonically treated samples displayed a higher degree of crystallinity (~37%) than the untreated ones (~31%) ($p \leq 0.05$) and a small peak at ~21° from crystallized saturated fatty acids. According to this, sonication promoted the complexation of FAs, in agreement with the results found by ATR-FTIR and with a previous work [3]. This finding may be attributed to the cavitation effect produced by ultrasound that disrupts the starch granules favoring the release of amylose and also improves the FA dispersibility in the starch solution, which increases the amylose–FA interaction to form ICs [3].

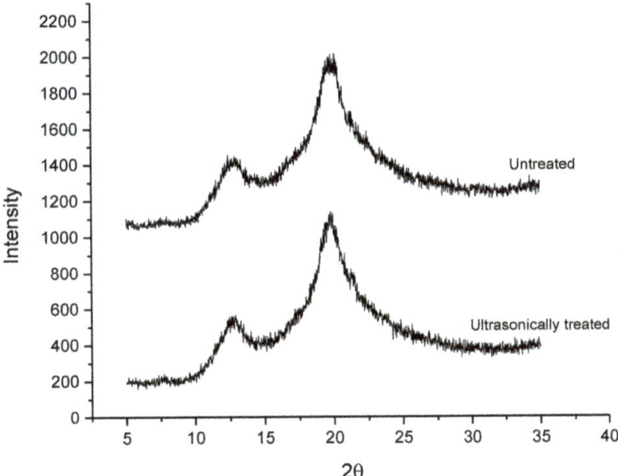

Figure 3. X-ray diffraction patterns of untreated and ultrasonically treated inclusion complexes of high-amylose corn starch with chia seed oil fatty acids.

3.4. *Differential Scanning Calorimetry*

Through DSC, it was observed that the ICs displayed a broad endothermic transition (Figure 4), which could be associated with the amylose–ligand dissociation. This confirms the effective complexation of chia oil fatty acids, in agreement with the XRD results. Both the untreated and ultrasonically treated ICs showed high-temperature stability (peak temperature (T_p) >90 °C) and a melting enthalpy (ΔH) of ~6 J/g (d. b.). No significant differences were found in the thermal parameters of ICs ($p \geq 0.05$), indicating that sonication did not have an effect on their thermal behavior. Since the ligand used in this work is formed by a mixture of fatty acids, the thermal transitions observed may be the result of several overlapped individual endotherms, thus yielding a broad dissociation range (Figure 4).

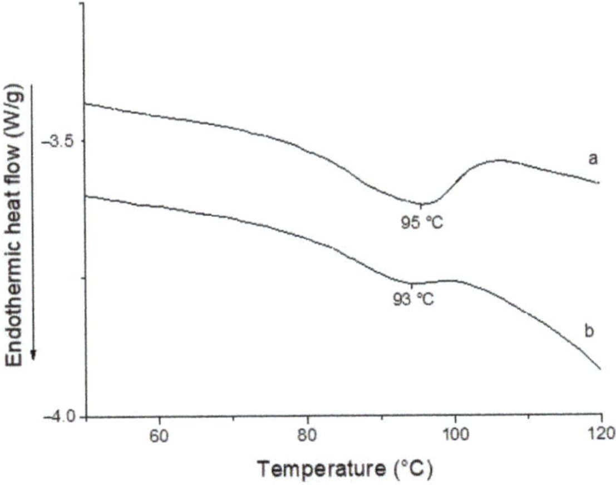

Figure 4. DSC thermograms of (a) untreated and (b) ultrasonically treated inclusion complexes of high-amylose corn starch with chia seed oil fatty acids.

4. Conclusions

From the results of the present work, it can be concluded that CR lipase had a better performance than PP to hydrolyze chia seed oil. The free fatty acids obtained after this reaction have demonstrated to successfully form V-type ICs with high-amylose corn starch, as verified by complementary XRD, DSC, and ATR-FTIR. The incorporation of ultrasonic treatment to the formation process promoted the complexation of these FAs, as inferred from the increased crystallinity and higher band intensity observed in the XRD and ATR-FTIR analysis, respectively. The high T_p of ICs (>90 °C) would suggest that this system may potentially act as vehicle of chia seed oil fatty acids in thermally treated foods.

Author Contributions: A.E.D.M.: Conceptualization, methodology, investigation, formal analysis, writing-original draft, visualization. V.Y.I.: Conceptualization, methodology, writing—review & editing, supervision. M.C.T.: Conceptualization, methodology, writing—review & editing, supervision, project administration, funding acquisition. All authors have read and agreed to the published version of the manuscript.

Funding: This work was supported by grant Ia ValSe-Food-CYTED (119RT0567), by the Universidad Nacional de La Plata (UNLP) (11/X907), by the Agencia Nacional de Promoción Científica y Tecnológica (PICT 2016-0323, 2020-01274), and by the Consejo Nacional de Investigaciones Científicas y Técnicas (PIP 2007).

Institutional Review Board Statement: Not applicable.

Informed Consent Statement: Not applicable.

Data Availability Statement: Not applicable.

Acknowledgments: Authors wish to thank Ingredion Inc. for the donation of high-amylose corn starch and Daniel Poire (CIG) for his technical assistance.

Conflicts of Interest: The authors declare no conflict of interest.

References

1. Di Marco, A.E.; Ixtaina, V.Y.; Tomás, M.C. Analytical and technological aspects of amylose inclusion complexes for potential applications in functional foods. *Food Biosci.* **2022**, *47*, 101625. [CrossRef]
2. Chao, C.; Yu, J.; Wang, S.; Copeland, L.; Wang, S. Mechanisms Underlying the Formation of Complexes between Maize Starch and Lipids. *J. Agric. Food Chem.* **2018**, *66*, 272–278. [CrossRef] [PubMed]
3. Kang, X.; Liu, P.; Gao, W.; Wu, Z.; Yu, B.; Wang, R.; Cui, B.; Qiu, L.; Sun, C. Preparation of starch-lipid complex by ultrasonication and its film forming capacity. *Food Hydrocoll.* **2020**, *99*, 105340. [CrossRef]
4. Di Marco, A.E.; Ixtaina, V.Y.; Tomás, M.C. Inclusion complexes of high amylose corn starch with essential fatty acids from chia seed oil as potential delivery systems in food. *Food Hydrocoll.* **2020**, *108*, 106030. [CrossRef]
5. Mendes, A.A.; Oliveira, P.C.; De Castro, H.F. Properties and biotechnological applications of porcine pancreatic lipase. *J. Mol. Catal. B Enzym.* **2012**, *78*, 119–134. [CrossRef]

biology and life sciences forum

Proceeding Paper

Variations in the Composition of "Algarrobas" (*Prosopis* sp.) Flours from Paraguayan Chaco †

Rocio Villalba, Juan Denis Ibars, Karen Martínez, Eva Coronel, Adeline Friesen and Laura Mereles *

Facultad de Ciencias Químicas, Universidad Nacional de Asunción, San Lorenzo P.O. Box 1055, Paraguay
* Correspondence: lauramereles@qui.una.py

† Presented at the IV Conference Ia ValSe-Food CYTED and VII Symposium Chia-Link, La Plata and Jujuy, Argentina, 14–18 November 2022.

Abstract: *Prosopis alba* and *Prosopis chilensis*, popularly called carob trees in the South American Chaco, are arboreal species. Carob fruits are an ancestral food for human consumption, mainly in the form of flour. In recent years, the study of carob trees in Paraguay has been based on the development of silvo-pastoral systems for livestock or as animal feed; very little is known about the compositional characteristics of the different varieties of carob that are part of the food systems, and that are used for the production of flours. Samples of flour from three autochthonous varieties of carob trees from the Central Chaco are evaluated for human consumption as a potential food ingredient in processed foods. They are evaluated for nutritional contribution, antioxidant potential and the preliminary evaluation of safety at the microbiological level. Official AOAC methods were used. The carob flour samples presented low humidity (less than 6%) and water activity (less than 0.45). The flours of the three species analyzed presented significant differences in their content of carbohydrates, lipids, proteins, dietary fiber and, consequently, in their caloric value, with a high content of polyphenols and antioxidant potential detected by ABTS. Presence of mesophilic aerobes, total coliforms and yeasts in the samples was observed. These results demonstrate the great food potential of carob flour from the Paraguayan Chaco, and indicate the need to address the food safety aspects of this type of wild-harvested food, to enhance their added value as ingredients for foodstuffs in the diet of regional populations.

Keywords: *Prosopis alba*; *Prosopis chilensis*; algarrobas; composition; flour; autochthon foods; Paraguay

Citation: Villalba, R.; Denis Ibars, J.; Martínez, K.; Coronel, E.; Friesen, A.; Mereles, L. Variations in the Composition of "Algarrobas" (*Prosopis* sp.) Flours from Paraguayan Chaco. *Biol. Life Sci. Forum* **2022**, *17*, 25. https://doi.org/10.3390/blsf2022017025

Academic Editors: Norma Sammán, Mabel Cristina Tomás, Loreto Muñoz and Claudia Monika Haros

Published: 16 November 2022

1. Introduction

It has been reported that Prosopis pods can constitute food sources from indigenous systems, such as "patay, arrope, chicha, aloja" in South America in the Gran Chaco region, where the species of Prosopis are native [1]. The genus Prosopis belongs to the Leguminosaseae family; its pods are used to obtain a harina, used in an ancestral way by native peoples, which has been shown to have a high nutritional value. The species of *Prosopis* are known by the name of "algarrobos" in the South American Chaco and are used as food, fodder, fertilizer, wood and raw material for the development of various economic activities. Very little is known about the nutritional characteristics of the different varieties that form part of the food systems and are used for the creation of *Prosopis* sp. flour in Paraguay. In the last decades, the importance of these ancestral foods in food security and the sustainability of ecosystems has been recognized [2]. Vulnerable conditions in rural populations, such as food insecurity and malnutrition, motivate the realization of development projects that are sustainable, and that implement alternative raw materials as a food base. In this context, the research of some South American countries have recently included foods such as "algarrobo" flours, which can vary in nutritional composition depending on the species of *Prosopis*, the form of obtaining it and the storage conditions [3]. The aim of this work was to evaluate the nutritional characteristics, the antioxidant potential and the presence of microorganisms in *Prosopis* sp. flour samples from an autochthonous species of "algarrobos" from the Central Chaco as a potential food ingredient in processed foods.

2. Materials and Methods

2.1. Sampling

Prosopis pods were manually collected from wild trees at Filadelfia, Boquerón, Chaco, Paraguay. They were recognized and stored by a local company and processed to obtain the flours. The three flour samples were made from *Prosopis alba* (white), *Prosopis chilensis* (yellow) and *Prosopis chilensis* (brown) pods (Figure 1) without seeds in successive stages of washing, drying, milling, sieving and potting. The *Prosopis* flour samples were packaged independently, and stored in dark multi-laminated aluminum bags until they arrived at the laboratory.

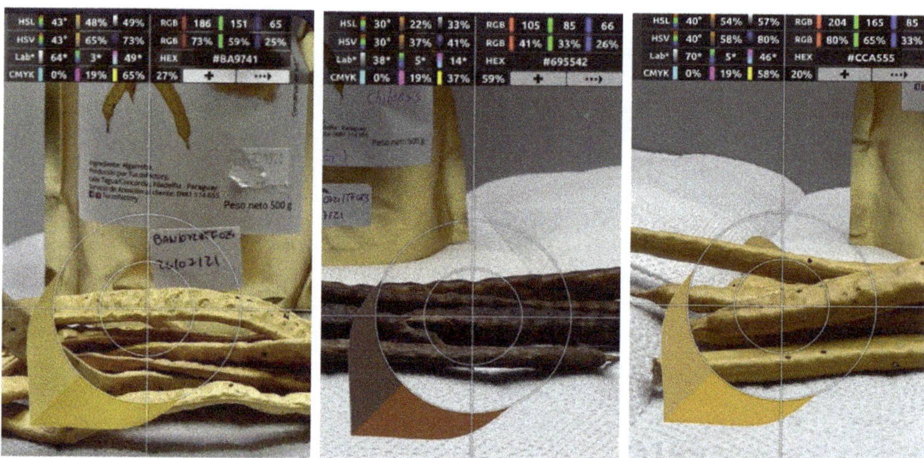

Figure 1. Representation of the pod colors of *P. alba* (white), *P. chilensis* (brown), and *P. chilensis* (yellow).

2.2. Analytical Methods

Prosopis flours were analyzed by official methods as follows; moisture AOCS Ca 2b–08, ash AOAC 900.2, proteins AOAC 979.09, total lipids AOAC 948.22, dietary fiber AOAC 985.29, total carbohydrates by anthrone method, water activity AOAC 978.19, and the caloric value by the Atwater method. Total phenols were determined using the Folin–Ciocalteau method with some modifications based on a colorimetric oxide-reduction reaction [4]. Total Antioxidant Capacity (TAC) by ABTS+ radical cation bleaching assay [5]. The microbial load was determined by Mesophilic aerobic AFNOR 3M 01/01-09/89, total coliforms AFNOR 3M 01/01-09/89, fecal coliforms AFNOR 3M-01/5-03/97 B, Yeasts and Mold count A.O.A.C. 997.02 and were used for analysis.

2.3. Statistical Analysis

The data was recorded and processed in a form of the GraphPad Prism 8.2 program (GraphPad Software Inc., San Diego, CA, USA). To determine significant differences, ANOVA and Tukey's post-hoc test was carried out and $p \leq 0.05$ was considered significant.

3. Results and Discussion

3.1. Physicochemical Characteristics and Antioxidants

Prosopis flour samples presented low moisture (less than 6%) and water activity (less than 0.45). The water activities of samples were: *Prosopis alba* (white) 0.393 $\pm$ 003; *Prosopis chilensis* (brown) 0.433 $\pm$ 006; and *Prosopis chilensis* (yellow) 0.378 $\pm$ 0.09. The samples analyzed presented a high content of carbohydrates (48.33–58.13 g/100 g) and total dietary fiber (25.67–32.15 g/100 g) in their composition, with significant differences (ANOVA, Tuckey post test $p \leq 0.05$) in their content of carbohydrates, lipids, proteins, dietary fiber and, consequently, in their caloric value (Table 1). Galera et al. [6] reported higher carbohydrate contents (72.47 g/100 g for *Prosopis chilensis* and 66.69 g/100 g for *Prosopis alba*) in the same species collected in Argentina, while studies carried out in Paraguay [7], Brazil and Bolivia [8] reported similar values of carbohydrates to those observed in this work for samples of same species of *Prosopis* flours (40.3–40.6 g/100 g *Prosopis alba*, 43.8–44.5 g/100 g *Prosopis chilensis*). Sucrose constituted the main sugar for flours obtained from *Prosopis alba* and *Prosopis nigra* [1]. The total lipid content in samples was low, which coincides with other regional studies [7,8] where the lipid content does not exceed 1.7%.

Table 1. Physicochemical characteristics of *Prosopis* sp. flours.

Compounds	*Prosopis alba* (White)	*Prosopis chilensis* (Yellow)	*Prosopis chilensis* (Brown)
Moisture (g/100 g)	$5.62 \pm 0.22\ ^{a}$	$4.31 \pm 0.31\ ^{b}$	$4.55 \pm 0.37\ ^{b}$
Ash (g/100 g)	$5.46 \pm 0.15\ ^{a}$	$4.21 \pm 0.10\ ^{c}$	$4.86 \pm 0.18\ ^{b}$
Total lipids (g/100 g)	$1.94 \pm 0.18\ ^{a}$	$2.79 \pm 0.10\ ^{b}$	$2.22 \pm 0.22\ ^{a}$
Total proteins (g/100 g)	$7.31 \pm 0.42\ ^{a}$	$9.41 \pm 0.44\ ^{c}$	$10.70 \pm 0.25\ ^{b}$
Dietary fiber (g/100 g)	$31.06 \pm 2.65\ ^{a}$	$25.67 \pm 1.81\ ^{b}$	$32.15 \pm 3.47\ ^{a}$
Carbohydrates (g/100 g)	$48.33 \pm 3.38\ ^{a}$	$58.13 \pm 1.55\ ^{b}$	$53.37 \pm 8.19\ ^{ab}$
Caloric value (kcal/100 g)	$240.01 \pm 14.8\ ^{a}$	$295.27 \pm 7.0\ ^{b}$	$276.26 \pm 9.1\ ^{b}$

Results are expressed as mean ± SD of three independent assays. Values in the same row with the same superscript letter are not significantly different ($p > 0.05$) as measured by ANOVA and Tukey's post-hoc test, $p \leq 0.05$.

Around 2.9% and 1.4% of soluble proteins were reported, for *P. alba* [1]. The caloric value ranged from 240 to 295 kcal/100 g with significant differences between the means (ANOVA, Tuckey post test, $p < 0.05$). The *Prosopis alba* flour had the lowest caloric value.

The TPC content on *Prosopis* flours samples were 610 ± 31 mg GAE/100 g in *P. alba* (white), 835 ± 82 mg GAE/100 g in *Prosopis chilensis* (brown) and 746 ± 18 mg GAE/100 g in *Prosopis chilensis* (yellow). The samples presented a good antioxidant potential: *P. alba* (white) 21.8 ± 4.07 mM TEAC/g, *P. chilensis* (yellow) 21.6 ± 0.68 mM TEAC/g and *P. chilensis* (brown) 23.1 ± 1.99 mM TEAC/g. It has been reported that the antioxidant potential of the Prosopis pods is greater in the dark pods such as *P. nigra* than in the cases of light-colored pods such as *P. alba* [1], which coincides with our findings.

3.2. Microbiological Analysis

Regarding the microbiological analysis of the samples, the results show the absence of fecal coliforms and fungi (Table 2). At the local level, there are no regulations on quality criteria for raw materials or Prosopis flour. However, the quantified levels of yeasts and Mesophilic Aerobes indicate the need to address the safety aspects of this type of wild food in future studies, to improve their added value as food ingredients in the diet of regional populations. These results demonstrate the nutritional potential of the analyzed samples, and suggest evaluating safety criteria in current production, to enhance their added value as food ingredients in the diet of regional populations.

Table 2. Microbiological analysis of *Prosopis* sp. flours.

	Recuento de Colonias (UFC/g)				
Flour Samples	**Mesophilic Aerobes**	**Total Coliforms**	**Fecal Coliforms**	**Molds**	**Yeasts**
P. alba (white)	6.8×10^3	4.4×10^3	-	$\leq 10^1$	6.0×10^3
P. chilensis (brown)	8.8×10^3	5.0×10^2	-	$\leq 10^1$	4.3×10^3
P. chilensis (yellow)	7.5×10^3	6.1×10^2	-	$\leq 10^1$	5.0×10^3

4. Conclusions

The flours obtained from *Prosopis* pods of different species from the same region can vary in composition of macro and micro components, which can affect the nutritional quality. Therefore, different species of *Prosopis* can present hybrids with different composition. Going forward, with good manufacturing practices, *Prosopis* flours could help to prevent pathologies associated with oxidative stress because they are a non-conventional source of antioxidant compounds.

Author Contributions: Conceptualization, L.M. and R.V.; methodology, K.M.; validation, J.D.I. and K.M.; formal analysis, R.V.; research, J.D.I.; Resources, A.F.; data curation, L.M.; writing—original draft preparation, J.D.I.; writing—reviewing and editing, E.C.; display, L.M.; supervision, L.M.; project management, L.M.; money acquisition, L.M. All authors have read and agreed to the published version of the manuscript.

Funding: This work was supported by grant "Project For Our Great Sustainable Chaco: Active participation in territorial management models for environmental conservation integrated with sustainable production", executed by COOPI-Cooperazione Internazionale in consortium with CERDET and FCQ-UNA, and the European Union "Conservation, sustainable use and good governance of biodiversity in four vulnerable biomes in central South America" (EuropeAid/154653/DD/ ACT/Multi).

Institutional Review Board Statement: Not applicable.

Informed Consent Statement: Not applicable.

Data Availability Statement: Not applicable.

Acknowledgments: The authors wish to thank Ia ValSe-Food-CYTED (119RT0567). Facultad de Ciencias Químicas and Tucos Factory E.I.R.L.

Conflicts of Interest: The authors declare no conflict of interest.

References

1. Cardozo, M.L.; Ordoñez, R.M.; Zampini, I.C.; Cuello, A.S.; Dibenedetto, G.; Isla, M.I. Evaluation of Antioxidant Capacity, Genotoxicity and Polyphenol Content of Non Conventional Foods: Prosopis Flour. *Food Res. Int.* **2010**, *43*, 1505–1510. [CrossRef]
2. Merchant, E.V.; Simon, J.E. Traditional and Indigenous Foods for Food Security and Sovereignty. In *Reference Module in Food Science*; Elsevier: Amsterdam, The Netherlands, 2022.
3. Cattaneo, F.; Costamagna, M.S.; Zampini, I.C.; Sayago, J.; Alberto, M.R.; Chamorro, V.; Pazos, A.; Thomas-Valdés, S.; Schmeda-Hirschmann, G.; Isla, M.I. Flour from Prosopis Alba Cotyledons: A Natural Source of Nutrient and Bioactive Phytochemicals. *Food Chem.* **2016**, *208*, 89–96. [CrossRef] [PubMed]
4. Singleton, V.L.; Orthofer, R.; Lamuela-Raventós, R.M. Analysis of Total Phenols and Other Oxidation substrates and antioxidants by means of folin-ciocalteu reagent. In *Methods in Enzymology*; Academic Press: Cambridge, MA, USA, 1999; Volume 299, pp. 152–178.
5. Re, R.; Pellegrini, N.; Proteggente, A.; Pannala, A.; Yang, M.; Rice-Evans, C. Antioxidant Activity Applying an Improved ABTS Radical Cation Decolorization Assay. *Free Radic. Biol. Med.* **1999**, *26*, 1231–1237. [CrossRef]
6. Galera, F.M.; Food and Agriculture Organization; Plant Production and Protection Division; Universidad Nacional de Córdoba (Argentina). *Las Especies del Género Prosopis (algarrobos) de América Latina con Especial Énfasis en Aquellas de Interés Económico*; Universidad Nacional de Córdoba: Córdoba, Argentina, 2000. (In Spanish)

7. Michajluk, J.; Mereles, L.; Wiszovaty, L.; Piris, P.; Caballero, S. Evaluación preliminar y del valor nutricional de vainas de Prosopis alba y Prosopis chilensis cosechadas en comunidades indígenas del Dpto. Boquerón, Chaco. *Rojasiana.* **2009**, *8*, 59–64. (In Spanish)
8. González Galán, A.; Corrêa, A.D.; Abreu, C.M.P.; de Fatima, M.; Barcelos, P. Caracterización química de la harina del fruto de Prosopis spp. procedente de Bolivia y Brasil. *Arch. Latinoam. Nutr.* **2008**, *58*, 309–315. (In Spanish)

MDPI
St. Alban-Anlage 66
4052 Basel
Switzerland
www.mdpi.com

Biology and Life Sciences Forum Editorial Office
E-mail: blsf@mdpi.com
www.mdpi.com/journal/blsf

www.ingramcontent.com/pod-product-compliance
Lightning Source LLC
LaVergne TN
LVHW070849170726
843515LV00005B/606

* 9 7 8 3 0 3 6 5 8 8 3 6 0 *